THE SOCIAL
CONTEXT OF
INNOVATION

THE SOCIAL CONTEXT OF INNOVATION

Bureaucrats, Families, and Heroes
in the Early Industrial Revolution,
as Foreseen in Bacon's *New Atlantis*

by Anthony F. C. Wallace

PRINCETON UNIVERSITY PRESS · PRINCETON, NEW JERSEY

Published by Princeton University Press, 41 William Street,
Princeton, New Jersey
In the United Kingdom: Princeton University Press, Guildford, Surrey

Library of Congress Cataloging in Publication Data will be
found on the last printed page of this book

This book has been composed in Linotron Trump

Clothbound editions of Princeton University Press books
are printed on acid-free paper, and binding materials are
chosen for strength and durability

Printed in the United States of America by Princeton
University Press, Princeton, New Jersey

Designed by Laury A. Egan

CONTENTS

LIST OF ILLUSTRATIONS

PREFACE

In the course of an earlier study of Rockdale, an American cotton manufacturing district in the early nineteenth century, my interest was attracted to the fraternity of mechanicians who invented the machinery of the Industrial Revolution. I conceived of this as a small cadre of men, on the order of a few hundred, who over the course of about three hundred years (roughly 1600 to 1900) made the great mechanical improvements that preceded the electrical age. A Guggenheim Fellowship (1978-1979) provided the leisure to explore the subject further, and this book is part of the result. At first it was planned as a kind of parallel to Robert Merton's classic collective biography of the scientists of the seventeenth century. But the collation of what were often sparse life-histories did not seem to answer the eventual purpose, which was to study not so much the attributes of individual inventors, but rather their mode of organization and support. Once this was defined as the problem, the development of the steam engine and of coke-smelted iron—central to the rise of a high-energy culture—suggested themselves as exemplary case histories, for which an extensive secondary literature and published collections of primary materials were available. In the meantime, having begun a study of the mining town of St. Clair as a companion-piece to *Rockdale*, the relative failure of technological innovation in some parts of the American coal industry promised a kind of control study of conditions under which the march of mechanical progress seemed to halt. After arranging these pieces, it dawned on me that that prescient administrator Francis Bacon had at the beginning of the Industrial Revolution tried to describe the research and development institutions of the future in an essay entitled *New Atlantis*. And hence the title of the book.

Sir Francis Bacon (1561-1625), Baron Verulam, Viscount St. Albans, was not a scientist nor an inventor. He was a lawyer who pursued a successful career in the legal departments of the British royal households from Elizabeth to Charles I, rising eventually to the highest post of all, Lord Chancellor. From this eminence he fell, near the end of his life, after an increasingly anti-royalist Parliament found him guilty of accepting bribes (a charge to which he confessed). His avocation, however, was the amassing of the great library at his estate in the country outside London, and the writing of a series of philosophical works that were famous even in his own time. These treatises were earnest pleas for a reorientation of the European intellect, and of European colleges and schools, away from its current preoccupation with sterile classical models of science and literature and toward the empirical study of nature on a large scale. Bacon's notions of how to go about scientific investigation by the inductive method may not satisfy modern canons of research design, but the importance of his call to build a new England on the basis of new empirical knowledge cannot be underestimated. It popularized science at the very moment when Galileo, Harvey, and Copernicus were launching the ship of modern science, and inspired the men who were to build the new technology of the future world.

MY THANKS go to Princeton University and to Herbert Bailey, the Director of the Princeton University Press, for the invitation to deliver the 1981 Stanford Little Lectures, which gave me the occasion to write the three essays that follow, and to Gail Filion, the Press's social science editor; to Eugene Ferguson and Brooke Hindle, who read the manuscript and contributed valuable suggestions for its improvement; to the Guggenheim Foundation and to the University of Pennsylvania, for financial support during

my fellowship year; and to the librarians and their staffs
at the libraries and historical societies where I have worked,
particularly at the Eleutherian Mills Historical Library,
the Historical Society of Pennsylvania, the Schuylkill
County Historical Society, the Van Pelt Library at the
University of Pennsylvania, and the Library of the American Philosophical Society.

Much of the material, particularly the anthracite mining material, has been presented in lectures and seminars
and has benefited greatly from criticisms and suggestions
from students and colleagues. I owe a particular debt of
gratitude to Barbara Finkelstein, whose invitation to read
a paper to the American Educational Studies Association
spurred me to consider the industrialist-as-hero theme as
it related to the problem of coal-mine safety. Alan Trachtenberg at Yale, Merritt Roe Smith at the Massachusetts
Institute of Technology, and Barbara Tedlock at Tufts,
and the students at colloquia at the University of Pennsylvania, Columbia University, and Wayne State University have all given me a chance to expand on these and
related topics. I owe much to my patient and skilled research assistants on the project, Dora Horchler, Denise
Schwartz, Teresa Cataldi, Adrienne Kols, Wendy Pollock,
and Pam Crabtree (who among other duties searched patiently through the personnel records of the Office of Ordnance in an unsuccessful effort to find Captain Thomas
Savery on the payroll in the 1690s). And finally, I owe a
particular debt to my wife Betty who has spent hundreds
of hours reading and copying microfilm of the Pottsville
Miners' Journal—a job of mining in itself!

PHILADELPHIA
MAY 1981

THE SOCIAL
CONTEXT OF
INNOVATION

INTRODUCTION

THE THREE ESSAYS constituting this book are concerned with the invention and introduction of new technology in the early Industrial Revolution. As the allusion of the title to Bacon's technological utopia suggests, however, our attention will focus on the institutional context of technological innovation rather than the technology itself. We shall view technology as a social product and shall not be over much interested in the priority claims of individual inventors, for the actual course of work that leads to the conception and use of new technology *always* involves a group that has worked for a considerable period of time on the basic idea before success is achieved. Thus the direction and speed of technological innovation is inevitably affected by the institutional setting, even if the concatenation of ideas in any one domain proceeds by an inner logic of its own that, like the development of science, is not explained by sociological factors. We shall be offering a social—more precisely, an institutional—interpretation of events already well known to scholars and will therefore provide not so much the results of primary research as the restudy of readily available data.

The purpose of these essays, then is to contribute to the aspect of the history of technology that views technology as a social product and examines the organizations in which innovators did their work, the structures of communities, trades, and workshops in which specific technological innovations were being invented and introduced. A prime example of this approach is Merritt Roe Smith's *Harper's Ferry Armory and the New Technology*.[1] Three other, more traditional, approaches have been es-

[1] Smith 1977.

3

sential in the development of an intellectually significant history of technology: first, and most essential, the analysis of the development of the technological ideas themselves, which can be performed with only minimal reference to personalities or societies (an excellent example is Robert S. Woodbury's series of histories of machine tools[2]); second, the biographical, whether by means of individual biographies (such as Samuel Smiles's famous *Lives of the Engineers*[3]) or by a kind of group biography (such as is, in part, Robert Merton's well-known study of the religious affiliations of seventeenth-century English scientists and technicians[4]); and third, the functional, that is, the study of the social consequences of technological innovation (one thinks here, for instance, of Lynn White's *Medieval Technology and Social Change*[5]).

No study of the social context of the early Industrial Revolution can avoid dealing with the issue of religion. A long tradition of scholarly inquiry has sought to relate the moral values and theology associated with the Protestant Reformation to the burgeoning of capitalist enterprise, the rise of science, and the proliferation of new technology. Of particular relevance to our inquiry is the work of Robert Merton and Charles Webster. Merton in *Science, Technology and Society in Seventeenth-Century England* showed that a disproportionate number of English and continental scientists and mechanicians were of Protestant persuasions.[6] Webster's *The Great Instauration* demonstrated the dependence of the Baconian tradition (culminating in the formation of the Royal Society) upon the Puritan millennialist vision of the Book of the prophet Daniel. Daniel foresaw a rebirth of learning and

[2] Woodbury 1972.
[3] Smiles 1966.
[4] Merton 1970.
[5] White 1962.
[6] Merton 1970.

a completion of man's dominion over nature.[7] If one looks at the inventors who collaborated to produce the great innovations, however, one finds a mixture of religious types—Catholic, Protestant, and (eventually) Deist. The names salient in the early history of steam—Giambattista della Porta, Salomon de Caus, Cornelius Drebbel, the Marquis of Worcester and Caspar Kalthoff, Sir Samuel Morland, Denis Papin, Thomas Savery, Thomas Newcomen and John Calley—reveal no particularly Protestant or Puritan bent. Della Porta was Catholic and so was the Marquis of Worcester. Drebbel was a Mennonite and Papin a Huguenot; both were vehement in their faiths. Morland was a somewhat unsteady Puritan who turned Anglican with the Restoration of his patron Charles II, and Savery would probably have been a loyal Anglican from the start. Newcomen was a Baptist and so probably was his partner Calley. Concerning the religious affiliation of de Caus and Kalthoff we have no information. In any case, the common thread that unites these men is more their consuming interest in the solution of technological problems than any personal religious inspiration. The fact of a general support of science and technology by both Catholic and Protestant states, furthermore, cannot be questioned, and in England the Anglican Crown as well as the Puritan Parliament both before, during, and after the Civil War believed in making at least a modest investment of public funds in scientific and technological research and development, along lines generally proposed by Francis Bacon.

Religious values had another role to play that helps to account for England's priority in the development of steam. France, her most likely competitor in the seventeenth century, like Germany in the sixteenth was hampered in the effective support of scientific and technological progress by religious factionalism. It was due specifically to

[7] Webster 1975. For a critical review of Webster's thesis see Mulligan 1980.

the persecution of the Huguenots that Papin, like many other French Protestants, found refuge and did much of his work in England. Neither Anglicans nor Puritans in England were as extreme as the French in their condemnation of non-conformists. And it is as much because of the Anglican monarchs' casual support of dissenting innovators as to Puritan technological millenarianism that the nourishment of the steam engine project was continued for so long.

In the case of the use of coke in the iron industry, again the prehistory of the enterprise shows a mixture of Anglicans and non-conformists cooperating, albeit unsuccessfully. Success eventually came to the Darbys of Coalbrookdale, a Quaker family; but the iron trade had Anglican as well as non-conformist families, who ran their businesses in much the same way as the Darbys, and it is difficult to conclude that membership in the Society of Friends was a necessary pre-condition to ingenuity in substituting coke for charcoal.

An even more specific difference between British and French state administration also helps to account for the British priority. It is not so much that the excessively bureaucratized French system, which as Nef has pointed out contrasted so sharply with the loose British administrative style, prevented middle-class capitalists in France from innovating as freely as their counterparts in England.[8] After all, in England, the early experiments in steam up to and including Savery's were not conducted by private capitalists. Rather, the French administration's central weakness was that it *had* virtually no ordnance establishment for most of the seventeenth century. Despite the efforts of Colbert, the French iron industry was never able to compete with England's in the production of suitable iron and in the casting of iron cannon. France depended heavily on the iron industries of other countries to supply her needs for artillery. England, by contrast,

[8] Nef 1940.

6

maintained the largest and most sophisticated ordnance establishment in Europe. It was furthermore a loosely administered establishment, embracing both private gun foundries (particularly those of the Browne family at Benchley) and publicly supported armories and administrative offices at the Tower, gun wharves (and later an arsenal and laboratory) at Woolwich and elsewhere, and research and development facilities at the Minories and—most importantly for the development of steam—at Vauxhall in Lambeth. It provided niches in out-of-the-way places for unobtrusive experimental projects of all sorts that might some day (or might never) produce a gadget useful to the Office of Ordnance, to the navy, or to the realm as a whole. There, with modest support, a variety of odd inventors could try their luck at new torpedoes, breech-loading guns, machine guns, submarines, pumps—and steam engines. There, too, they could train, learn from, and maintain in residence a cadre of mechanicians and workmen, both native and foreign, ready to undertake other new projects as assigned. All of this could be, and perhaps was, deliberately done with minimal fanfare, lest ordnance establishments other than Britain's learn of British developments before they became visible in the field.

This was not quite the way Francis Bacon had planned it, perhaps, when he first envisioned a college for inventors, with an "Inginary" and a Library, in 1608. He had imagined a state guided by a college of philosophers, rather than a school for artisans maintained by the Master-General of Ordnance. But Bacon was a worldly man as well as a philosopher; and, toward the end of the passage on Salomon's House, as he described the gallery wherein were placed "the statues of all principal inventors," he listed third, after Columbus and the inventor of ships, the English "monk who was the inventor of ordnance and of gunpowder."[9] That monk's surname, of course, was also Bacon.

[9] Bacon, *Works*, 1: 351-413 ("New Atlantis").

Thus it would seem that while Protestantism in most of its forms found the hard work of studying nature for the sake of private or public good a highly congenial occupation, a commitment to the support of science and technological improvement was already state policy in most European countries, Catholic as well as Protestant. We must look to the specific institutional arrangements that facilitated—or inhibited—technological innovation in order to find a more exact answer to the question: How did the great inventions take place?

The three essays that follow take up this challenge. In the first, we shall see how the British Royal Ordnance and naval establishments, which maintained staff and facilities for experimental weaponry, unintentionally nourished the development of the steam engine in a process lasting over one hundred years. In the second, we shall look at the Darby connection of Coalbrookdale, whose family firm in three successive management generations introduced and nearly completed the transition from charcoal to coke in making iron. From the examples of steam, iron, and coal we may conclude that the necessary institutional setting for the major innovations of the Industrial Revolution was a stable transgenerational organization that provided continuity in both plant and personnel so that a technological idea could become paradigmatic in a single community of mechanicians. Sir Francis Bacon's utopian essay *New Atlantis* expressed that insight 350 years ago.

The third paper, however, describes an industry unable to make use of the available technological innovations necessary for its own survival: deep shaft coal mining, as it was practiced by British emigrants following British methods in the middle of the nineteenth century in the anthracite districts of Pennsylvania. The short life span of a coal mine (twenty years or so) and the small profit per ton of coal mined (five to ten cents) meant that only short-term planning was attractive to most operators; social approval rewarded daring entrepreneurs even for their failures. In this setting, safety procedures, plans, and devices, old as well as new, were cast aside in the rush for wealth and fame. Safety did not become a paradigm to the operators of coal mines even though failure to accept its central economic importance led invariably to their own bankruptcies.

I

The Royal Office of Ordnance and the Powers of Steam and Air

THE PROTOTYPE of all modern research and development institutions was Francis Bacon's dream of Salomon's House or College of the Six Days Works, on the island of New Atlantis. In his posthumously published essay entitled *New Atlantis*, the eloquent administrator set forth his image of the kind of institution that would most effectively encourage scientific and technological advance.

SALOMON'S HOUSE

The idea of a research and development institution supported by the state was not original with Francis Bacon. He may have known of the Ptolemies' Museum at Alexandria, with its library and its laboratories for scientific and technological study and experiment; he almost certainly was aware of Tycho Brahe's observatory, Uraniborg, and of the allegedly mad Emperor Rudolf's assembly of scientists and mechanicians in Prague.

Bacon himself began to think seriously about a great institute for technological and scientific research and development as early as 1608, when he made notes on the founding of a "college for Inventors" to contain "a Library and an Inginary . . . Vaults, fornaces, Tarraces for Isolacion; woork houses of all sorts."[1] The institutions that

[1] Colie 1955, 246.

11

he cited as administrative (not intellectual) models were Trinity College at Cambridge (his own alma mater) and Magdalen at Oxford. Twelve years later, in the preface to dedication of *The Great Instauration*—in which he described again his design for an intellectual rebirth of mankind and a renewal of man's empire over nature, the "great instauration" foreseen by the prophet Daniel—he urged James I to support the development of the new universal science, based on observation and experiment, with royal patronage.[2] This college was to be a department of state; one might almost say that the King was to be an agent of the college. And so it remained in the final form of Bacon's thinking on the subject, published posthumously in 1627 as an imperfect fragment—the utopian fable of the *New Atlantis*, the kingdom ruled by knowledge.[3]

The New Atlantis was a small group of islands in the South Seas that had been substantially isolated from the rest of the world ever since the collapse of the once great civilizations of the Old World, and later of the Americas. That catastrophe had ended the first age of remote voyages. Such wandering seafarers as were cast upon the New Atlantis's shores almost invariably elected to remain; and the research expeditions regularly sent out from the island to gather the best of mankind's knowledge invariably managed to return undetected. Bensalem, as this antipodal kingdom was named, remained a self-sufficient archipelago, undisturbed by internecine wars or by the fluctuations of the world economy, a haven of Christian virtue and unceasing progress in a decadent world.

The laws and institutions of Bensalem had been established by a King Solamona, who reigned three hundred years before the birth of Christ, and who became (in an anthropologist's jargon) the culture hero of the island. He it was who instituted the policy of isolation, in the in-

[2] Bacon, *Works*, 1: 23-24.
[3] Bacon, *Works*, 1: 351-413. Quotations from the *New Atlantis* are taken from this edition.

terest, essentially, of preventing change for the worse. Bacon's language is illuminating on the risks of social change and the unrestricted diffusion of ideas:

> and recalling into his memory the happy and flourishing estate wherein this land then was, so as it might be a thousand ways altered to the worse, but scarce any one way to the better; thought nothing wanted to his noble and heroical intentions, but only (as far as human foresight might reach) to give perpetuity to that which was in his time so happily established. Therefore amongst his other fundamental laws of this kingdom, he did ordain the interdicts and prohibitions which we have touching entrance of strangers.

Despite the generally conservative bent of his design, Solamona also created (and evidently established a permanent endowment to support) the Society of Salomon's House, "which house or college . . . is the very eye of this kingdom." Salomon's House occupied a unique place in Bensalem; as one of its wise men explained:

> Ye shall understand (my dear friends) that amongst the excellent acts of that king, one above all hath the pre-eminence. It was the erection and institution of an Order or Society which we call *Salomon's House*; the noblest foundation (as we think) that was ever upon the earth; and the lanthorn of this kingdom. It is dedicated to the study of the Works and Creatures of God.

Salomon's House, also called the College of the Six Days Works, had the dual purpose of improving both science and technology:

> The End of our Foundation is the knowledge of Causes, and secret motions of things; and the enlargement of

the bounds of Human Empire, to the effecting of all things possible.

Bacon composed the *New Atlantis* toward the end of a life largely devoted to service as a public official. Despite some criticism of his grasp of political realities, he was no stranger to problems of administration. He was a conservative, protective of the claims of the Crown against the pretentions of a restless and increasingly middle-class Parliament; he believed in an established church as the guardian of public morals; and he sought to develop a technologically progressive nation ruled by a wise and learned monarch. He had served as an active member of the House of Commons from 1584 to 1618; trained in the law, he had served the Crown as Solicitor-General, Attorney-General, and for three years as Lord Chancellor. His writings were devoted in considerable part to reform of the universities, to pruning away the dead branches of classical studies, and to the provision of adequately equipped laboratories and libraries, to be used by adequately paid professors. Although only an amateur of science and technology, Bacon was a professional in administration, and his conception of a large, independently endowed research institution embodied the astute observation that dependence on the vagaries of royal patronage by individual philosophers or mechanics would not ensure systematic development of knowledge.[4]

It is also worth observing that Salomon's House did not propose investigation in what we now call the social sciences. Bensalem was a perfect and unchanging society, undisturbed by vice or crime, steady in its devotion to the principles established by King Solamona two thousand

---

[4] Bacon's vision of the role of science in an ideal kingdom, and its influence on the social policy of the Puritans in the period 1626-1660, has been described in detail in Webster 1975. Among the several critical evaluations of Bacon as philosopher, one that sees him most clearly as a "political engineer" is Paterson 1973.

years before. Its most important institution (apart from Salomon's House, of course) was the patriarchal extended family, in whose honor a special, royally licensed ritual was performed when any man should live to see thirty of his descendants assembled together. Strict monogamy was required, as it was in More's *Utopia* (which Bacon cites), and, in glaring contrast to dissolute Europe, both males and females normally remained faithful to their spouses for life. Bensalem was "the virgin of the world." But it was also a strict monarchy, and Bacon makes no mention of a Parliament; innovation was not to be social but scientific and technological. What need was there for a Parliament, let alone a science of society, when all laws were just and perfect, every man worked at his trade, the king was a rational and learned man, and Salomon's House constantly improved the material conveniences of life?

The resources and organization of the Society of Salomon's House were extensive. There was, to begin with, a large physical plant—underground laboratories and mines, some of them over half a mile deep (a depth not quite achieved in the seventeenth century even in the deepest gold and silver mines of Central Europe); experimental and observation stations on high towers, some of these reaching a height of half a mile; lakes, pools, wells, baths, and fountains; great buildings with chambers for all sorts of physical, chemical, and biological research; orchards, gardens, parks, and zoos; breweries, bakeries, kitchens, and pharmacies; furnaces and foundries, optical and accoustical laboratories, and engine houses (where improved cannon, submarines, aircraft, clocks, and "perpetual motions" were made). Bacon's recitation of the conventional interests of natural and experimental philosophers of his day does not overshadow the magnitude of his conception. To endow the physical establishment he describes would tax the resources of a major university, a National Science Foundation, a Manhattan Project, or a space program of our own time; in the early seventeenth century, it would require the concentrated wealth of a kingdom.

A complete table of organization is not provided, and estimate of the numbers of laborers and artisans is not even attempted. Bacon simply notes that there are novices and apprentices (who are the students who eventually became members of the Society) and "a great number of servants and attendants, men and women." But Bacon does give an outline of the division of labor among the supervisory personnel. There were thirty-six senior investigators in the Society:

> we have twelve that sail into foreign countries, under the names of other nations (for our own we conceal); who bring us the books, and abstracts, and patterns of experiments of all other parts. These we call Merchants of Light.
>
> We have three that collect the experiments which are in all books. These we call Depredators.
>
> We have three that collect the experiments of all mechanical arts; and also of liberal sciences; and also of practices which are not brought into arts. These we call Mystery-men.
>
> We have three that try new experiments, such as themselves think good. These we call Pioneers or Miners.
>
> We have three that draw the experiments of the former four into titles and tables, to give the better light for the drawing of observations and axioms out of them. These we call Compilers.
>
> We have three that bend themselves, looking into the experiments of their fellows, and cast about how to draw out of them things of use and practice for man's life, and knowledge as well for works as for plain demonstration of causes, means of natural divinations, and the easy and clear discovery of the vir-

tues and parts of bodies. These we call Dowry-men or Benefactors.

Then after divers meetings and consults of our whole number, to consider of the former labours and collections, we have three that take care, out of them, to direct new experiments, of a higher light, more penetrating into nature than the former. These we call Lamps.

We have three others that do execute the experiments so directed, and report them. These we call Inoculators.

Lastly, we have three that raise the former discoveries by experiments into greater observations, axioms, and aphorisms. These we call Interpreters of Nature.

Although Bacon is not explicit on the point, an intellectual hierarchy is implicit in the inductive procedure revealed in his description of the roles of the senior investigators. That such an arrangement of scientific roles describes neither modern science nor scientific organization is commonly acknowledged; but we have noted that Bacon was not so much a scientist, let along a technological innovator, as an administrator, educator, and scientific popularizer. He divided responsibility for the various aspects of a very large enterprise among teams or committees that administered their own sections, with coordination and synthesis achieved, rather democratically, in meetings of all members of the Society. Perhaps he was indeed thinking of a college faculty as the model, each department carrying out its own scholarly task in relative independence, but joining when a critique upon an effort at interdisciplinary synthesis was required. The whole Society also met on at least one other type of occasion: to deliberate upon the risks of innovation:

And this we do also: we have consultations, which of the inventions and experiences which we have dis-

17

covered shall be published, and which not: and take all an oath of secrecy, for the concealing of those which we think fit to keep secret: though some of these we do reveal sometimes to the state, and some not.

Those "profitable inventions" that met with the Society's approval were announced to the citizens of the principal cities of the kingdom during periodic visits by representatives of the Society, along with warnings and prognostications of earthquakes, tempests, plagues, and other natural disasters.

Bacon's conception of Salomon's House reveals two principles of organization. In the recitation of the physical layout, Bacon assigned particular structures to appropriate subject-matter specialties, or what we might today call disciplines. Thus the "high towers" were reserved for meteorology (the study of "winds, rain, snow, hail"); the "dispensatories or shops of medicine" for the preparation of traditional and synthetic drugs were assigned to pharmacology; a special house was set aside for mathematics and mathematical and other scientific instruments; and so on. But in each of these fields the order of research and development proceeded inductively from data-collection to generalization with different persons being assigned to the tasks of each level, under the general supervision of the thirty-six senior officers, who somehow coordinated the various disciplines. Since, in practice, much of the study would take the form of the preparation of an encyclopedia of "natural histories" (of everything from the artisans' trades to the departments of nature), there may have been a certain argument for such an arrangement on the ground that the methodologies of each discipline would be basically the same on every level of the inductive process.

For the mechanical part of technology, Salomon's House made provision in several of its apartments for the investigation of engines of various kinds to produce, convert, and distribute powers and forces. There were devices for harnessing the energies of "violent streams and cataracts" and for "multiplying and enforcing the winds." Bacon does not directly admit that Salomon's House was experimenting with water wheels or windmills, however; in his discussion of deep mines, he makes no reference to the problem of pumps (which had been discussed at length in Agricola's classic treatise on mining and metallurgy, *De Re Metallica*, some seventy years before and with which Bacon was familiar). This neglect of an avowed commitment to the advancement of mining and metallurgical technology is curious because Bacon had invested in tin mines in Cornwall and a wire-drawing ironworks at Tintern. In addition, he had been personally involved in the administration of the patent law under James I, sitting as one of the commissioners and of necessity being officially cognizant of each application for a patent of monopoly (including inventions). Most patent applications involved mining and metallurgy and many concerned the problem of drainage by pumps. Bacon himself was an advocate of the use of tunnels, or adits, for draining mines.[5] Yet the *New Atlantis* does not mention these domains of technological innovation. The reason may be that innovations in these spheres were commonly and anonymously accomplished in the course of practical work by miners, smelters, and other tradesmen. Although Bacon respected and proposed "natural histories" of the mechanical trades, the trades themselves were not organized to permit their incorporation into the collegiate model of a research and development institution, being at that time arranged as chartered companies with guild-like features.

The central institution for mechanical research and innovation in Salomon's House was the "Inginary" (as he

---

[5] Webster 1975, 345-346, reviews Bacon's connection with the patent office and his personal involvement in the technology of mines and wireworks.

19

termed it in 1608, when he first committed his dream of a "college for inventors" to writing). His description in the *New Atlantis* is contained in a single paragraph.

We have also engine-houses, where are prepared engines and instruments for all sorts of motions. There we imitate and practise to make swifter motions than any you have, either out of your muskets or any engine that you have; and to make them and multiply them more easily, and with small force, by wheels and other means: and to make them stronger, and more violent than yours are; exceeding your greatest cannons and basilisks. We represent also ordnance and instruments of war, and engines of all kinds: and likewise new mixtures and compositions of gun-powder, wildfires burning in water, and unquenchable. Also fire-works of all variety both for pleasure and use. We imitate also flights of birds, we have some degrees of flying in the air; we have ships and boats for going under water, and brooking of seas; also swimming-girdles and supporters. We have divers curious clocks, and other like motions of return, and some perpetual motions. We imitate also motions of living creatures, by images of men, beasts, birds, fishes, and serpents. We have also a great number of other various motions, strange for equality, fineness, and subtilty.

The sources for Bacon's conception of an "inginary" were no doubt diverse. Rosalie Colie has shown that the work of two of his contemporaries, Salomon de Caus and Cornelius Drebbel, may well have provided the model for a number of the domains of Atlantean experiment, including meteorology, artificial fountains, optical instruments, visual illusions, and clocks.[6] Of the claims asserted in the "engine-houses" paragraph, Drebbel is almost certainly

[6] Colie 1955.

responsible for the mention of the "perpetual motion" and of the submarine, which he successfully demonstrated by a three-hour immersion in the Thames in 1620 before an astonished crowd of courtiers and city-folk, and which is said to have successfully traveled from Westminster to Greenwich. Drebbel's submarine, indeed, is said to have been a tourists' attraction at dockside for some twenty years. Bacon's reference to experiments in flight may reflect knowledge of Leonardo's ideas, or of the actual flights of Eilmer of Malmsbury, an eleventh-century Benedictine monk whose exploits were well known in England in the seventeenth century, or of the speculations of many contemporary philosophers of nature.

Interesting though it may be to pursue the sources of all of Bacon's mechanical allusions, our purpose is to consider the social arrangements that he was proposing for the development of science and technology. For mechanical technology, his image of a suitable setting was not the artisan's workshop nor the gentleman's laboratory nor philosopher's cabinet, but the "Inginary"—an institution of a very different kind.

NAVAL GUNNERY AND THE OFFICE OF ORDNANCE

The Inginary or engine-house (or in more modern language "works" or "shop") was a familiar scene in early modern Europe on a small scale in armories, mines, foundries, and dockyards. At dockyards naval vessels were constructed, outfitted, and armed with brass or, in England, iron cannon. These activities required mechanical aids in the form of cranes, pulleys, and carriages for the handling of bulky materials and heavy objects. The royal dockyard at Woolwich on the Thames, a few miles below Westminster, was particularly prominent because there the navy's largest ships were constructed and launched under the superintendency of the Cambridge-trained master shipwright Phineas Pett (1570-1647). Pett had studied mathematics

21

and drawing, had spent time at sea, and brought to the task of designing ships a sophisticated understanding of naval architecture and a command of the logistics involved in operating a major dockyard, with its attendant gun-wharf, rope-walk, and storage sheds and shops. He built and launched at Woolwich not only the largest British ships of war, the famous *Prince Royal* (1610) and the *Sovereign of the Seas* (1637), but also some of the largest merchant ships.[7]

During this same period, another naval experiment—to which Bacon alludes—was being conducted by the famous Dutch Mennonite mechanician and alchemist, Cornelius Drebbel (1572-1633). Drebbel in his youth was apprenticed to an engraver, in whose home he lived, and eventually married one of his master's younger sisters—a woman of beauty, expensive tastes, and, according to family tradition, relentless lubricity. His mechanical talents were not confined to the engraver's art, for he began to experiment with all sorts of devices—pumps, clocks with a "perpetual motion," chimneys, fountains. James I was interested in the perpetual motion clock, which seems to have been an elaborate astronomical calendar operated by diurnal variations in temperature or air pressure, and he invited Drebbel to England. About 1605 Drebbel moved from the Netherlands to London and was set up in the service of Henry, the Prince of Wales, with living quarters and laboratory space at Elpham Castle.

Elpham Castle had been a royal residence from the fourteenth century to the time of Henry VIII; but that monarch built a new palace at nearby Greenwich, and the Palace was only occasionally visited by Elizabeth and James I. The castle was particularly famous for its great hall, which survives to this day, but most of the other structures fell into disuse and ruin after being damaged during the Civil War. Elpham stood on a hill overlooking Green-

---

[7] DNB "Pett, Phineas"; Haas 1980, 419-428.

wich (then the Ordnance Department's powder magazine and later the site of the royal observatory) and Woolwich dockyard and gun-wharf (and later the major arsenal), a mile or two downstream. Drebbel left Elpham for Prague in 1610 and the court of Emperor Rudolf II, then the declining monarch of the Holy Roman Empire. There Rudolf had assembled a notable group of scientists and mechanicians, including Tycho Brahe and Johannes Kepler; but Rudolf died in 1612, and, after some difficulty, Drebbel returned to Elpham.

In 1620 he demonstrated a submarine, very likely constructed at Woolwich, before the King and his court. According to an eyewitness, it remained submerged—with no funnel or hose connecting it to the surface—for three hours. Another source, published in 1645, asserted that the ship had been rowed under water from Westminster to Greenwich, a distance of six miles. The secret of the crew's ability to breathe during prolonged submersion apparently lay in the fact that Drebbel, an able empirical chemist, had discovered how to generate oxygen from the heating of saltpetre and to store it in bottles that were opened as the crew required; Robert Boyle credited him with having recognized that air was composed of different gases of which one was necessary to support life. Drebbel also engaged in a variety of other innovative enterprises, including the construction of thermostats for furnaces, the making of compound microscopes, the development of naval torpedoes, a scheme for draining swampy land, and the marketing of a process for making scarlet dye (named for his son-in-law Abraham Kuffler and made famous by the Gobelins).

Drebbel's contribution to the development of the steam engine, however, was made before he left the Low Countries. In 1604 he published, in Dutch, a treatise on the *Elements of Nature*, which contained an explicit account of experiments in which water was sucked up into a cooled retort from which either air or steam had been driven off by previous heating. The account was illustrated by a clear diagram. Drebbel thus demonstrated the power of atmospheric pressure to drive standing water into a vessel containing a partial vacuum. This was one of the two main technological ideas on which the later development of the steam engine depended. The other idea—the power of expanding steam to drive either a column of water or a piston out of a vessel—was already known from the work of della Porta and Salomon de Caus, whom we shall discuss shortly. Drebbel's book was popular and, being translated into French and Latin, was certainly accessible to educated Englishmen.[8]

Drebbel's involvement with naval weaponry reminds us of another aspect of the Woolwich complex, the gunwharf where cast-iron guns were landed from the ships bearing them from the foundries and where the ships of the line were outfitted with ordnance. The gun-wharf was administered by the Royal Ordnance Department whose headquarters were in the Tower of London, not far upstream. The principal duty of the Ordnance Department at this date was to contract for iron cannon, test them at

1. Cornelius Drebbel's Atmospheric-Pressure Pump (1621)

[8] For Drebbel's biography see Tierie 1932, Rye 1865, and Harris 1961; an account of Elpham Palace is given in Buckler 1828.

24

the Tower, store them, and outfit the royal navy and army with them. Inasmuch as a large proportion of the realm's iron was mined, smelted, and cast into guns in the Weald, a forest in Kent and Sussex about forty miles south of London, the south bank of the Thames at London and its forested hinterland constituted, in the early seventeenth century, a major concentration of military and naval technology, rich in examples of the enginery to which Bacon was alluding, and all more or less coordinated by the Royal Ordnance Department.

The Office of Ordnance was established as a separate department of the state early in the fifteenth century, and its budget was next in size to that of the Treasury itself. The Master of Ordnance and his principal lieutenants were appointed by royal patent and by the end of the sixteenth century (when Bacon's patron and friend the Earl of Essex was, among other duties, Master of the Ordnance) they administered an extensive physical establishment and a staff in London and elsewhere that must have included thousands of men. At the accession of James I, the title of Master of Ordnance was changed to Master General of Ordnance. In 1622 the position was described in part as follows:

As thus the Master of the Ordnance hath these commandments in remote, foreign, and out-of-the-way places: so hath he in the camp as eminent and great controlments; for there the general charge of the whole artillery dependeth upon him and his necessary substitutes, of which the principal are the Lieutenant of the Ordnance, the Clerk of the Ordnance, the paymaster, the purveyor-general, four scribes, four stewards, a harbinger, a chancellor, divers interpreters, a chaplain, a physician, a surgeon, a trumpeter, all engineers and refiners, a guard of both horse and foot, gentlemen of the ordnance and halberdiers, and over all these several places . . . he is the chief superintendent, and hath the power to dispose of all things according to his pleasure and judgment, as also he hath the command, choice and controlment of all gunners and cannoniers whatsoever, and both giveth onto them their several allowances, and doth allot them their several attendants.

It is also in the power of the Master of the Ordnance to press and have under him both shipwrights, boat-wrights and other necessary carpenters, who at his appointment shall frame boats, barges and other vessels which may be profitable, and at pleasure taken assunder and joined, for the transportation and carriage of the army over any great rivers, or small arms of the sea, by fastening these boats together, and making bridges thereof . . . as hath been done in divers foreign armies, and also with us here at home in the year eighty-eight, when the army and provisions were ferried over the Thames between Kent and Essex: so that of these boats for bridges should never be in the army under the number of forty at least, over which charge (under the Master of the Ordnance) should be a captain of the boats, two shipwrights, a master-carpenter to plank them, twenty sailors and caulkers, a guard of horsemen to conduct them, two smiths and their men to have charge of the ironwork; a master of the cables, anchors and graplings, a wheelwright and certain carters to drive the carriages.

The Master of the Ordnance appointeth under his signature the numbers and proportions of all manner of munitions which shall attend the army, and delivereth to the lieutenant who seeth them provided and distributed to the inferior officers, and the inferior officers keep them in their charge, and dispose or deliver them out as they shall receive warrant either from the Master of the Ordnance, or his lieutenant.

Under the command of the Master of the Ordnance, is the carriage-master, the clerk of the carriages, the harbinger, the steward, the gilmaster, a provost, two carpenters, two farriers and all the carters, horses, oxen and all that draw any kind of armament; . . .[9]

In the London area, the Office of Ordnance occupied the Tower, where its central offices were located, and where firearms, powder, cannon, armor, and other military and naval supplies were tested, repaired, and stored, awaiting distribution to military units and naval vessels. It had its own gun-wharf on the Thames. All cannon purchased by the Crown were proved by test-firing in an adjacent artillery garden. The main ordnance storehouse was the Minories, also adjacent to the Tower. In the period when he was working on the torpedo project, Drebbel was assigned quarters and workspace in the Minories. The Ordnance Department also managed the country's main powder magazine at Greenwich, and the gun-wharves at Woolwich and the other royal dockyards. Although the cannon that came into its hands were cast by the private gunfounders of the Weald (principally John and Thomas Browne in Kent), the Ordnance wrote the specifications and exercised quality control.

In addition to its prime mission in supplying the army and navy with cannonry, the Office of Ordnance was responsible for the design and fabrication of various military engines of the kind so carefully illustrated in the contemporary machine books—cranes, devices for mechanically hurling projectiles (such as incendiary bombs), gun carriages, and, of course, portable boats, landing craft, and pontoon bridges for spanning streams. The office was in charge of the design, armament, and maintenance of all major permanent fortifications at home and abroad. Furthermore, the office was responsible for the movement of all guns and munitions and thus had to maintain large

9 Quoted in Hogg 1963, 59-60.

numbers of horses and oxen, farriers, carters, and stable boys. The office thus combined many of the functions of the modern corps of artillery, transport, and engineers in addition to strictly ordnance functions.

The whole system was, however, extremely loose administratively; the Crown's style of management was gentlemanly and casual. Not only were guns supplied by private gunfounders under royal patents (i.e., written grants of monopolistic privilege to provide the Crown with a specified service), but the same system applied also to all the senior officials in the Office of Ordnance itself, each of whom had a separate lifetime patent from the Crown outlining a sphere of privileges and service within the general mission of the Office. These spheres of privilege and service sometimes overlapped. Hierarchical authority was thus extremely difficult for the Master General to maintain because all of his lieutenants could claim that their own authority derived from a royal patent and not from the senior officer. Thus, despite some technical control of founders by the Office of Ordnance on the proving ground, the office and the gunfounders were really coordinated both by the Crown itself, through the language of patents, and by Parliament, through laws controlling the making and disposition of guns. Disputes over the boundaries between areas of responsibility were consequently very difficult to resolve. The pay was low and malfeasance not uncommon. Nevertheless, as we shall see, it was in the administrative morass that constituted England's Office of Ordnance that the first serious efforts to develop the steam engine were begun.[10]

Bacon must have been familiar with the whole British ordnance-shipyard-gunfounding complex. In describing the Inginary, however, he actually may have had in mind an even more famous concentration of state-supported and

[10] A detailed history of the Office of Ordnance is provided in Hogg 1963, which, however, is devoted largely to the Royal Arsenal at Woolwich.

28

administered war-making technology—the celebrated arsenal of Venice. It was, in the early seventeenth century,
past its peak of production, but it was nonetheless still
so impressive a concentration of dockyards, gun-wharves,
gunfoundries, gunpowderworks, and storage sheds, re-
quiring elaborate mechanical apparatus for the transfer
and placement of heavy objects, that Galileo at the opening of his *Discourses on Two New Sciences* paid tribute
to it as a source of his own inspiration.

The constant activity which you Venetians display
in your famous arsenal suggests to the studious mind
a large field for investigation, especially that part of
the work which involves mechanics; for in this department all types of instruments and machines are
constantly being constructed by many artisans, among
whom there must be some who, partly by inherited
experience and partly by their own observations, have
become highly expert and clever in explanation.[11]

Later in the century, the English diarist Evelyn visited the
arsenal and marveled at its size and richness:

The arsenal is thought to be one of the best-furnished
in the world. We entered by a strong port, always
guarded, and, ascending a spacious gallery, saw arms
of back, breast, and head, for many thousands; in another were saddles; over them, ensigns taken from
the Turks. Another hall is for the meeting of the Sen-
ate; passing a graff, are the smiths' forges, where they
are continually employed on anchors and iron work.
Near it is a well of fresh water, which they impute
to two rhinoceros's horns which they say lie in it,
and will preserve it from ever being empoisoned. Then
we came to where the carpenters were building their
magazines of oars, masts, &c., for an hundred galleys
and ships, which have all their apparel and furniture

---

[11] Galileo 1939 (1638), 131.

29

near them. Then the foundry, where they cast ordnance; the forge is 450 paces long, and one of them has thirteen furnaces. There is one cannon, weighing 16,573 lbs., cast whilst Henry the Third dined, and put into a galley built, rigged, and fitted for launching within that time. They have also arms for twelve galeasses, which are vessels to row, of almost 150 feet long, and thirty wide, not counting prow or poop, and contain twenty-eight banks of oars, each seven men, and to carry 1300 men, with three masts. In another, a magazine for fifty galleys, and place for some hundreds more. Here stands the Bucentaur, with a most ample deck, and so contrived that the slaves are not seen, having on the poop a throne for the Doge to sit, when he goes in triumph to espouse the Adriatic. Here is also a galley of 200 yards long for cables, and above that a magazine of hemp. Opposite these, are the saltpetre houses, and a large row of cells, or houses, to protect their galleys from the weather. Over the gate, as we go out, is a room full of great and small guns, some of which discharge six times at once. Then, there is a court full of cannon, bullets, chains, grapples, grenadoes, &c., and over that arms for 800,000 men, and by themselves arms for 400, taken from some that were in a plot against the State; together with weapons of offence and defence for sixty-two ships; thirty-two pieces of ordnance, on carriages taken from the Turks, and one prodigious mortar-piece. In a word, it is not to be reckoned up what this large place contains of this sort. There were now twenty-three galleys, and four galley-grossi, of 100 oars of a side. The whole arsenal is walled about, and may be in compass about three miles, with twelve towers for the watch, besides that the sea environs it. The workmen, who are ordinarily 500, march out in military order, and every evening receive their pay through a small hole in the gate where the governor lives.[12]

A great deal more is known about the Venetian arsenal than about many comparable, if smaller, institutions because its records have survived. It was founded in 1104 and reached its zenith in the mid-sixteenth century; it is considered to have been "the foundation of the power of Venice, the heart of the state" (a phrase reminiscent of Bacon's references to Salomon's House as "the lanthorn of this kingdom" and "the very eye of this kingdom"). The Arsenal then covered sixty acres and employed within its walls about 1,500 workers, most of them belonging to special Arsenal-affiliated guilds; it was administered by senior officers in the naval administration, who held office for specific terms of years, and the workmen constituted a kind of tightly disciplined civil service. In effect, the Arsenal of Venice concentrated into one state-supported and bureaucratically managed sixty-acre tract what in England was distributed much more widely throughout 500 square miles of Thames River dockyards and armories and Wealden gun foundries, under a mixed administration of the naval and ordnance departments and private enterprise. If Bacon were choosing between Britain and Venice for the administrative model for Salomon's House, he undoubtedly, with his penchant for central administration, would have chosen the Arsenal.[13]

PUMPS FOR ENGLISH MINES AND WATERWORKS

The other major industrial domain that made use of heavy enginery was mining. Mining in England, technologically,

---

[12] Evelyn 1889, 1: 214-215.
[13] For accounts of the Arsenal as it was in the seventeenth century, see Rapp 1976 and Pullan 1968; for earlier periods, Lane 1934 (who referred to it as the "foundation of the power of Venice, the heart of the state," p. 152).

was as far behind continental, and particularly German, practice, as English gunnery was in advance, despite the ready availability of the illustrated machine books of Ramelli, Zonca, and Biringuccio, and of the classic text on mining and metallurgy, Agricola's *De Re Metallica*.[14] England mined little copper (a principal reason for the substitution of iron for bronze artillery) but large quantities of tin (particularly in Cornwall) and was beginning to develop the extensive bituminous coal deposits that provided a substitute fuel for wood. The increasing demand on European forests in general was denuding the landscape of trees in many places, for wood was used as the principal domestic fuel, as the source of charcoal for smelting and forging iron, as the source of fuel for all other heat-requiring industries as well, and for naval timber. The historian John U. Nef has even seen the sixteenth-century fuel crisis, and the resultant efforts to substitute coal for wood as the new energy source, as an industrial revolution of its own.[15]

As mines, whether metal or coal, became deeper, the technical problems of hoisting, ventilation, and water drainage became increasingly difficult. Putting aside for the moment the matters of hoisting and ventilation, let us turn to the problem of draining water out of coal mines. The digging of adits (slightly down-sloping drainage tunnels) was often impracticable because both veins and overlying topography were relatively flat. At the beginning of the seventeenth century the pits in English collieries were seldom deeper than sixty feet because adequate enginery was not available to raise water any higher. By this date, metal mines in Germany were raising water hundreds of feet using force pumps, which had been known from Roman times, and either horses or water-wheels to supply

[14] Agricola (Georg Bauer) 1950 (1556); Biringuccio 1942 (1540); Ramelli 1976 (1588); Zonca 1607.

[15] Nef 1966; see especially Ch. II "An Early Industrial Revolution," 1: 165-189.

the power, some of these operating the crank at the pit-head from hundreds of yards away by means of draw-rods. But none of this technology was in use in England, most collieries depending on adits or, when that was impossible, on the primitive "chain of buckets" and "rag and chain" systems.[16] Even under these conditions, some collieries were able to produce and sell on the order of 20,000 tons of coal per year, and could have produced even more if they had been able to go deeper. There was a virtually feverish interest in new pumps and other devices for mines. About 75 percent of the 317 or so patents for invention granted between 1561 and 1668 were directly or indirectly concerned with mining, and of these about 20 percent were for the drainage of mines.[17]

In addition to their role in mining, pumps were also required for two other kinds of enterprise of public interest. The drainages of the fens and other low-lying flood plains and swamps, particularly in the neighborhood of London; and the movement of water to cisterns and reservoirs where it could supply water under gravity pressure to castles, towns, and fountains (the latter a never-ending object of admiration to garden-loving Englishmen). Windmills, long used in the low countries to power the pumps, began to dot the English landscape where drainage projects were under way. Prominent engineers and philosophers, among them, Salomon de Caus and the ubiquitous Cornelius Drebbel, involved themselves in thinking about better pumps and the means to power them.

It was, in fact, in connection with the pumping problem that the second principle for using the power of steam to do mechanical work was articulated by an engineer resident in England. Born in France, Salomon de Caus (1576-

---

[16] For draw-rod technology, see Hollister-Short 1976, 1977; Multhauf 1959. Draw-rods have remained in use in Europe and America up to the present time, particularly in oil fields to work several pumps from a single power source.

[17] The figures are taken from Nef 1966, 1: 255.

1626) had been educated in mathematics and the writings of the classical engineers including Archimedes and Hero of Alexandria (who made devices involving the use of steam jets as a propulsive power). De Caus, after visiting Italy, came to England about 1608, where he was employed as a tutor in mathematics to the same Henry, the Prince of Wales, who was being served by Drebbel, and as drawing master to the Prince's sister Elizabeth. When Elizabeth married the German elector palatine Frederick V and moved to Heidelberg in 1615, de Caus accompanied her. During his seven years in England, however, he carried out experiments in the mechanical powers of water and was extensively employed in laying out the gardens

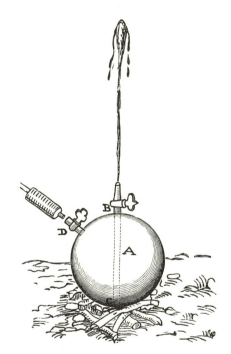

2. Salomon de Caus' Steam-Pressure Pump (1615)

34

at Greenwich and in other landscape architecture projects at the royal castles around London. In 1615 he published the work which is responsible for the preservation of his name in the annals of science, *Les Raisons des forces mouvantes avec diverse machines taut utilesque plaisantes—Ausquelles sont adjoints plusiers desseings des grotes et fontaines.* In this work he described a device, which he may or may not have actually built, that pumped water by the power of steam; it operated by principles previously illustrated by della Porta. Water confined in a copper vessel, laid on a fire, was partially transformed into steam, which of course collected at the top of the vessel. A pipe, closed by a valve, led up from the bottom of the vessel and passed through a sealed opening at the top; when the steam pressure was sufficiently high, the valve was opened and the high-pressure steam drove part of the water up the pipe. The valve was then closed, the fire removed, and after the container cooled, more water was added through a hole in the side of the vessel.[18]

The natural philosopher Giambattista della Porta had first described this method in an Italian work in 1601, and Cardano (whose father had been a friend of Leonardo, who had proposed a high-pressure steam gun) in the mid-sixteenth century had alluded both to the power of expanding steam and the vacuum obtained by condensing steam. Bacon was familiar with both writers, and the scholarly de Caus, who had traveled in Italy, undoubtedly knew della Porta's and Cardano's works too. Furthermore, steam had been used for some time to form a kind of substitute for the bellows, supplying warm air to medieval furnaces and forges. Bacon does not indicate any awareness of de Caus's specific proposal for raising water by fire, but at least one of his contemporaries seems to have done so, though it is not known whether his ideas were

[18] NBG, "Caus, Cauls, ou Caux (Salomon de)," 9: 258-260. DNB, De Caus, Cauls, or Caux, Salomon."   .

3. Giambattista della Porta's Steam-Pressure Pump (circa 1606)

attempts to apply de Caus's theorem. David Ramsay, who had been appointed in 1613 to the post of clockmaker-extraordinary to the King and who became the first master of the Clockmaker's Company, in 1630 received a patent "to raise water from low Pitts by fire . . . to raise water from low places and myres, and coal pits, by a new waie never yet in use" (along with a patent for a perpetual motion machine and a method for the manufacture of saltpetre so marvelous that he promised to supply the needs of entire kingdoms from four acres of ground!).[19]

Outside the great naval and ordnance system, enginery apparently remained for the most part in a primitive condition in England until long after Bacon's death. Effective steam pumps were not available for another three generations. Even deliberate efforts to introduce German metal mining technology failed, as in the case of the unsuccessful Cumberland copper mines, where the *Stangenkunst* or draw-rod system that transmitted power from water wheels to pit heads over hundreds of yards, was introduced by immigrant German engineers about 1600.

[19] DNB, "Ramsay, David."

36

The experiment was abandoned and never renewed; the mine drainage problem was not solved until the introduction of the steam engine a century later. The technological backwardness of English in comparison with German mining would seem to reside in the differences in institutional arrangement. In the German states the entire precious metal production system, from mining to smelting to minting, was under the control of the state. Elaborate laws and regulations, administered by royal officials, aimed to make sure that the Crown received its due and that the mines were operated with maximum efficiency. Innovations, once identified, were rapidly promulgated, for increased production was vitally important now that gold and silver from the Americas began to compete with that from European mines. The German prince was, in effect, a capitalist entrepreneur who leased the right to mine his gold and silver to highly organized mining companies. In England, by contrast, the King, who claimed a royal right only to gold and silver, preferred to tax, or even borrow from, small companies of adventurers who paid for patents of monopoly to mine other metals. Coal operators, too, dealt directly with the land owners and paid "royalties" to them rather than the Crown. The English enterprises thus, in this period, presumably aimed at quick profit with simple methods rather than long-term investment in expensive machinery. Furthermore, the companies of tradesmen and merchants, typical of England, and even more so the continental guilds were often outrightly hostile to new inventions, especially those patented, which threatened to provide unfair advantage to those practitioners of an art who were able to acquire the new device.

THE ORDNANCE WORKS AT VAUXHALL AND THE HIGH-PRESSURE STEAM PUMP

Our earliest evidence of an Ordnance Department policy of encouraging mechanical innovation was its interest in

37

Dud Dudley's coal-iron cannon. As early as 1612, the Crown began issuing patents for various processes for smelting iron with sea-coal or pit-coal. Those efforts were all less than successful, and the patents lapsed. Dud Dudley, the son of a Worcestershire ironmaster, saw someone else's experiments with pit-coal at Lambeth about 1615 and when he took over three of his father's furnaces in Worcestershire in 1619, he also received a pit-coal patent. His efforts were at least marginally successful and his guns were tried repeatedly at the Tower. Dudley later became an associate of David Ramsay, one of the earliest inventors to claim the fire-engine, and during the Civil War he served as general of the ordnance to the royalist Prince Maurice.[20]

The interest of the naval and ordnance departments in experimental naval weaponry was further shown, as we have seen, in the employment of Drebbel at the Minories next to the Tower from about 1620 on. That interest was again expressed in 1629 when the Crown acquired a house in Lambeth with several acres and a number of cottages. The place was turned over to the inventor of a leather-bound wrought-iron gun-barrel for the manufacture of these guns (and the guns were still being produced by his nephew in the building constructed there for the purpose as late as 1645). Shortly afterward, the Ordnance Department acquired the services of a Dutch engineer, Caspar Kalthoff, who was set up in the Tower with a workshop and a forge in 1633. About 1639 the works at Lambeth were enlarged by the purchase of an adjacent property, a soap factory and surrounding grounds, for the purpose of making a foundry. The agent on behalf of the Crown was an enthusiastic royalist named Edward Somerset, the son of one of the wealthiest lords in England, and soon to become the Marquis of Worcester. Kalthoff's operation was now shifted from the Tower to Vauxhall, as the Lambeth es-

---

[20] For accounts of Dudley's experiments with coal-iron, see Dudley 1665; see also Rovinson 1613, and Sturtevant 1612.

tablishment became known for its location on the ancient lands of the manor of that name.[21] It is clear that the Crown originally intended Vauxhall to be an experimental ordnance works. When the Baconian Puritan Samuel Hartlib informed his friend Robert Boyle that "Fauxhall is to be set apart for public uses, by which is meant making it a place of resort for artists, mechanics, &c. and a depot for models and philosophical apparatus . . . (where) experiments and trials of profitable inventions should be carried on . . . of great use to the Commonwealth," he added that the late King Charles I had "designed Fauxhall for such a purpose."[22] The Parliamentary plans for Vauxhall were detailed in a memorandum:

1. To keepe all manner of Ingenuities, rare Models and Engines which may bee useful for the Common-wealth.
2. To make Experiments and trials of profitable Inventions, which curious artists ofttimes cannot offer to the knowledge of skilful men and to public Use for want of a place of Adresse to meet with them, and of other necessarie conveniencies to show a proofe of their skill, whereof in Fauxhall is great store.
3. To bee a place of Resort wherunto Artists and Ingeniers from abroad and at home may repaire to meet with one another to conferre together and improve many ways their abilities, and hold forth profitable Inventions for the use of the Common-wealth.

THE REASONS
1. The late king did designe that place for such an Use and that wee should bee lesse mindful of the

[21] The history of the Vauxhall works is given in Jenkins 1971 (1936), Ch. 4 "The Vauxhall Ordnance Factory of King Charles I"; Thorp 1932-33.
[22] Dircks 1865, 266.

39

Public than hee did seeme to bee, will bee a disparagement unto us.

2. The conveniences of forges, furnaces, mills, and all manner of tooles for making of Models and Experiments being there already will bee a great losse to the Common-wealth if they should bee destroyed, and if the House bee alienated into some privat hand this will fall out.

3. In other Common-wealths as in Switzerland and the Low-Countries and the free Imperial Cities of Germany, there are places designed for all manner of Ingenuities, which they call *kunst-kameren* that is the chambers of Artifices.

4. It will encourage artistes of all sorts at home and abroad to looke towards us, to esteeme of us, and to repaire to us, as men of Public Spirits and Lovers of Ingenuities, which will not only bee a credit to the Parliament, but an occasion of much profit. For to have a magazin of all manner of Inventions, and a ready way to encrease the same and trie the usefulnes thereof, is a treasurie of infinit and unknow'n Value in a Common-wealth which by setting this place apart for Public Use may bee gained.[23]

Somerset and Kalthoff seem to have shared an interest in some sort of perpetual motion machine that raised water; Somerset even erected a model in the Tower in 1639. But Somerset was quickly caught up in the politics and then the actual conflicts of the Civil War, and by 1641 he had left the London area, now in the hands of Parliamentarians, for the safety of Raglan Castle in Monmouthshire. In 1642 he was denominated the Marquis of Worcester. Kalthoff fled to Holland, where in 1641 he obtained a patent for a machine whereby one pound of water would raise two or three. (In 1655 the Marquis of Worcester would claim credit, though not a patent, for a machine by means

[23] Quoted in Webster 1975, 364-365.

of which one pound would raise a hundred.) From 1642 until about 1646 the Marquis was engaged in the field on behalf of the King, and during that time made repeated efforts to raise a Catholic army in Ireland with the support of papal emissaries. With the collapse of the royalist forces in 1846, Worcester removed to France and remained there until 1851, when he returned to face trial (in the face of an earlier sentence of death passed by the House of Commons). He spent two years in the Tower. Upon his release in 1654 he leased or purchased Vauxhall and held it as "a school for artisans" until 1661, when it was returned to the Crown for the use of Caspar Kalthoff, who once again "was employed in the making of guns and divers engines and works for his Majesty's service." Worcester died in 1667 and Kalthoff in 1667 or 1668; but the Crown granted life tenure to most of the establishment, including the "Operatory," to Kalthoff's widow and his son-in-law (nominally a baker).[24]

In 1667 Sir Samuel Morland, a well-known mechanician of the time, moved into one of the Vauxhall houses; in 1675 he obtained a twenty-one-year lease of a major part of the tract, including house, workshops, and extensive gardens, and there employed "his mechanics" on a variety of projects, but particularly at elaborate pumps, fountains, and hydraulic systems. Morland died in 1695 and available records do not show what use was made of the Vauxhall establishment in the years immediately after his death. It is tempting to speculate that it was at Vauxhall, or at least with some of the mechanicians at Vauxhall, that Thomas Savery worked to produce his first steam engine in 1697 or 1698, for (according to a man who knew Savery) Savery demonstrated the first model "in a potter's house in Lambeth" in 1699, and there were potteries at Vauxhall about that time. By 1716 Vauxhall gardens were

---

[24] Dircks 1865 is the standard biography of the Marquis of Worcester and contains an extensive account of his activities at Vauxhall. See also Doorman 1947-49.

known as a pleasurable place to stroll and in 1732 the Vauxhall establishment was acquired by an entrepreneur who landscaped it as the famous New Spring Gardens or Vauxhall Gardens.[25]

The "works" at Vauxhall thus seem to have been maintained in more or less continuous operation as an "operatory" for experiments in various branches of engineering—chiefly ordnance, pumps, water works, and "perpetual motions"—from about 1629 until the turn of the century. It was located in a rural suburb of London that housed a number of highly skilled Dutch immigrant artisans, many of whom worked in the delft-ware potteries of Lambeth. Other Dutch artisans were specialists in brass and were employed both in the Vauxhall works and in nearby Wansworth in Lambeth, where Dutchmen manufactured brass plate for kettles, skillets, frying pans, and so forth, "and keep it as a mystery."[26] The equipment and organization of the Vauxhall works were described in 1645 in a detailed Parliamentary inventory of former Crown property and again in a site survey in 1668. It consisted of a narrow tract of land along the Thames about four hundred and fifty feet long and two hundred feet wide, walled in brick, and containing about two acres. There was one large house (Copped Hall, at the very edge of the Thames) where Kalthoff had lived and worked, and under his supervision an engineer, William Youlden, who had a royal patent or commission, and who actually made the models that Kalthoff designed. Copped Hall contained the great mill for the boring of guns and several rooms equipped with forges, bellows, anvils, vises, hammers, files, presses, benches, lathes (operated by bows), and miscellaneous other machinists' apparatus. The mill also contained two "Modell Roomes" containing models of lead, copper, wood, and

[25] General accounts of Vauxhall and the neighboring parts of Lambeth may be found in Aubrey 1719; Victoria County Histories of England, Surrey, 2: 281-285, 410-415; 4: 44-60; Walford n.d.: 447-467.

[26] Aubrey 1719, 14.

iron. The inventory provides evidence that research and development was the principal activity in Kalthoff's hall:

6. In the first Modell Roome two stories high, one leaden waterworke, seaven great wheeles made for a perpetuall mocion, one lettle copper waterworke, two waggon modells to carrie Ammunition in, one Modell for waggon to go without horses, one woodden moddell for a fortification, one woodden modell for an iron Mill to grind corne, foure water fountaines to stand in Chambers, one Modell of a waterworke to raise water, one other modell for the same use, one endless scrue of iron for a mocion worke of water, one long fixed table with one Shelfe, and divers other small tooles of iron and wood under the said Table, two perpetuall mocions whereof one of tinne plate, another of paistboard, the modell of a coach made of purpose to let loose the horses, if they should prove wild.

7. In the second Modell roome two stories high, 1. one great Table in the midst of the said roome whereupon there is one modell for the decke of a ship to catch and entrap men if they come aboard, 2. one modell for six guns to charge at the breach, 3. one modell of a boat to go of itself against the streame & tide, 4. one modell of a woodden lock, 5. foure Cannon Modells upon wheeles, 6. one little modell for a waterworke, 7. one woodden Modell for foure guns to charge at the breech, 8. one box for charging of six guns, 9. and lastlie, one modell in woodd for the working roomes.

On a second Table in this same roome, 1. one Modell of freestone to cast Saker Shott in, 2. one woodden Modell for a perpetuall mocion, 3. one Modell in paistboard for a gallerie, 4. one modell for a tomb or monum^t, 5. two woodden models for a round fortification, 6. one modell of a horne waterworks to caste water out of a trench, 7. & lastlie, one Modell for the fronte of a house or building. Yet, in the said second modell roome, the modell of a scaling ladder, the Modell of an Engine to cutt Tobacco upon, the Modell of a lether engine to cast water out of a Trench, one little crane to bring peeces of iron to the Anvill, one rest for a plaine Table, one other rest to Shoot of a musket, one Modell of wood for the breech of a great cannon, on both sides, one little frame of wood, two modells in wood for a perpetuall mocon, one little modell in paper for a fountaine, two little side Tables fixed to the wall, three shelves with divers peeces of wood & boards of no value.

In another structure, probably the former soap factory, was housed the ordnance shop, where brass cannon and various experimental firearms were made up, including breech-loading cannon, muskets and carbines, and arrangements of multiple guns mounted on wooden frames for rapid firing by one man. In a separate structure, the tile-roofed "melting house," were seven furnaces for melting the brass for casting cannon. Along the street was the brick house newly built for Colonel Robert Scott about 1629 for the manufacture of the leather guns, and still in possession of his nephew, Colonel Wemys. Finally, there was a long garden to the south with a "great butt" at the end at which the experimental artillery was fired to prove its soundness.[27]

In its heyday in the late 1630s, and perhaps again in the fifties and sixties, Vauxhall must have employed dozens of local craftsmen in its several departments of brass gunfounding, engine assembly, leather gun manufacture, and experimental model construction, together with numbers of household servants and grounds-keepers, all presumably on the payroll of the Office of Ordnance (but perhaps also with the financial assistance of the future Marquis

[27] Thorpe 1932-33.

of Worcester, who with his father was in the habit of advancing the financially straitened King large amounts of money).

The connection of the ordnance establishment in Lambeth with the development of the steam engine consists in the use of its facilities over a period of some sixty years by three men, and possibly four, who made experiments and published projections of (and in the last instance, produced) a steam engine for pumping water. These four men were Caspar Kalthoff, the Marquis of Worcester, Sir Samuel Morland, and (possibly) Thomas Savery. With respect to Worcester and Morland, it is true that, despite enthusiastic advocacy of their claims to priorities of one sort or another, no contemporary observer ever saw, or even alluded to anyone seeing, an actual demonstration of a fire engine pumping water. Worcester, however, in his famous *Century of Inventions* published in 1663 but allegedly written in 1655 on the basis of experiments conducted in collaboration with Kalthoff earlier at Vauxhall, did publish a reasonably clear description of a steam pump that alternately sucked water into, and ejected water from, each of two receivers, the action being manually controlled by two cocks:

An admirable and most forcible way to drive up water by fire, not by drawing or sucking it upwards, for this must be as the Philosopher calleth it, *Intra sphaeram activitatis,* which is but at such a distance. But this way hath no Bounder, if the Vessels be strong enough; for I have taken a piece of a whole Cannon, whereof the end was burst, and filled it three quarters full of water, stopping and scruing up the broken end; as also the Touch-hole; and making a constant fire under it, within 24. hours it burst and made a great crack: So that having a way to make my Vessels, so that they are strengthened by the force within them, and the one to fill after the other. I have seen the

45

water run like a constant Fountaine-stream forty foot high; one Vessel of water rarified by fire driveth up forty of cold water. And a man that tends the work is but to turn two Cocks, that one vessel of water being consumed, another begins to force and re-fill with cold water, and so successively, the fire being tended and kept constant, which the self-same Person may likewise abundantly perform in the interim between the necessity of turning the said Cocks.[28]

Kalthoff and Worcester had succeeded, at least in theory, in uniting Drebbel's and de Caus's two principles of condensation and expansion of steam. The action differs from their models, however, by separating the boiler from the receiver, so that energy and time need not be wasted in alternately heating and cooling a single vessel. Whether this apparatus is among those so ambiguously listed in the inventory of 1645 can probably never be known. There is no reason to suppose that Worcester and Kalthoff did not build and try their fire engine, nor that it would work at least briefly. But they were working with high steam pressures and much heat, the soldered joints were likely to melt or burst, steam and water to escape from leaky valves, and the receivers to be crushed by atmospheric pressure if too high a vacuum was created by the entry of cold water. Whence Worcester and Kalthoff derived their original inspiration is not known. An apocryphal story has it that Worcester met de Caus in a lunatic asylum in Paris during his exile, where the aged inventor, maddened by the indifference of Richelieu to his great discovery, imparted it to the English nobleman. De Caus, however, died in 1631, long before Worcester's exile; but it is conceivable that the young lord made the philosopher's acquaintance, or at least came to know his works, prior to 1625 during his student days when he acquired, as he later stated, a "virtuous education . . . in Germany, France, and Italy." In any case, Worcester and Kalthoff were able to

[28] Dircks 1865, 475-476.

46

achieve a measure of success with an improved force pump that enabled a single man to raise water to heights of forty feet or more with ease. Worcester's demonstrations of the pump (his "water-commanding engine") were successful (although his and Kalthoff's alleged perpetual motion earned them the contempt of the waspish Robert Hooke).[29]

The wonderful pump was still being demonstrated at Vauxhall after Worcester's and Kalthoff's deaths and during the initial days of Morland's tenancy. Sir Samuel was a more systematic thinker who attempted to quantify his mechanical experiments; he estimated the volume of steam as 2,000 times that of water at one atmospheric pressure—a figure not improved until the work of James Watt a century later. Anticipating Huygens, Morland was interested in the possibility of working a suction pump by exploding gunpowder in an enclosed chamber, the hot gases driving out much of the air and leaving a partial vacuum into which water would rush through a pipe thrust into the water below. He took out a patent on such a device, with a view to installing it in mines. His principal contribution to mechanics, however, lay in recognizing the inadequacy of the traditional practice of fixing the seal to the cylinder rather than the plunger of force pumps. He devised a means of attaching the leather seals to the plunger itself, thereby considerably increasing the efficiency of the pumps, reducing the wear, and vastly simplifying repair.

Although it has been argued that Morland devised a reciprocating high-pressure piston steam engine for use in driving rotary machinery by means of a crank attached to the piston, the idea would seem to be Savery's, to judge from a drawing set next to that of Savery's engine in Roger North's notebook.[30] Morland intended his force pumps

[29] See Dircks 1865, 292-293.

[30] Jenkins 1971 argues at length that Morland by 1682 had invented "a two-cylinder, single-acting, high-pressure condensing engine with automatic valve gear" (Ch. 8, "A Contribution to the History of the Steam Engine").

primarily for raising water to cisterns in the turrets of castles, palaces, and gentlemen's country seats, and he advertised and marketed them commercially. Morland's pumps were in fact so efficient that the weight of the columns of water they could raise exceeded the strength of the men operating them, and mechanical aids had to be applied. Part of his solution to this aspect of the problem was to counterbalance the weight of the pump rod, plunger, and water column by heavy weights hung from a beam, in an arrangement physically similar to Newcomen's beam-engine-and-pump of a later date. The weight of water also necessitated the construction of heavy lead pipes to resist the pressure. The motive power, he observed in his work *Elevation des eaux* (1685), could be anything at all—men, horses, water wheels, windmills, gunpowder, or "du feu ordinaire," presumably implying some sort of steam engine, the details of which he left unspecified. That Morland at Vauxhall was at work on fire-engines is clear enough, for he demonstrated in 1682 "a new invention of raising any quantity of water to any height by the help of fire alone." In December of that year a warrant was drawn up for the grant of a patent for the device, but it seems never to have been issued, and shortly thereafter Morland left for Paris hoping to see his pumps used to provide water for the royal palace at Versailles. In this he was disappointed but during his time in France he published his work on the *Elevation des eaux*, in which he stated the figure for the expansion of steam and talked about its utility for the raising of water.[31]

The successful completion of the Drebbel and de Caus paradigm of a steam engine was accomplished in Lambeth—and possibly at Vauxhall—in 1697 or 1698 by Captain Thomas Savery, the engineer (presumably an ordnance officer, civilian employee of the office who affected the title, or a naval officer concerned with engineering

---

[31] For a biography of Morland, see Dickinson 1970.

48

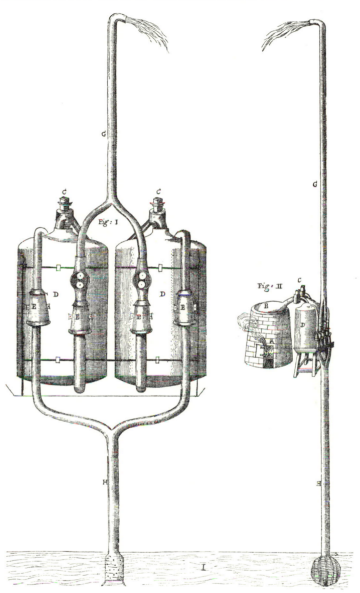

4. Thomas Savery's Steam-Pressure Engine (1699)

matters). Savery was a resolute inventor of all sorts of devices, including a number projected in Worcester's *Century* (and other philosophers' books of the time), and he translated into English a Dutch work on fortification that undoubtedly was of interest to officers of the Ordnance Department. Savery's engine, first constructed at "a Potter's House at *Lambeth*," was demonstrated at Hampton Court and then patented in 1698; it was described and illustrated in the *Philosophical Transactions* of the Royal Society in 1699; sketched by Roger North about 1701; and further described and pictured in Savery's own work, *The Miner's Friend* in 1702. Savery seems to have solved successfully the problems that Worcester and Kalthoff had been unable to master half a century before. Savery's engine embodied a pair of copper receivers that alternately sucked water from below in the vacuum phase and drove it upward in the high pressure steam phase, thus achieving a nearly constant flow of water in the wooden pipes. The copper boiler was built into a brick furnace and was recharged with water automatically. The brass control valves were self-acting. Indeed, the whole machine was a sophisticated device that required highly skilled workmen for the manufacture of the parts, their assembly at the site, and the management of the machine. Savery recommended his steam pump particularly for mines. There it had to be installed within twenty feet or so of the sump at the bottom of the pit (atmospheric pressure does not support a column of water even in a perfect vacuum any higher than thirty-two feet); but as Savery pointed out with considerable foresight, the draft of the fire would help to ventilate the mines. This method of mine ventilation continued in use long after Savery's engine was abandoned in favor of Newcomen's.[32]

[32] There is no substantial biography of Savery, who left few written remains beyond his patents and his publications (see Savery 1698, 1699, 1702). But see DNB "Thomas Savery" and Jenkins 1971, Ch. 9, "Savery, Newcomen and the Early History of the Steam Engine." Switzer 1724,

5. Thomas Savery's Steam-Pressure Engine (1702)

The significance of the Vauxhall establishment for the development of steam power is that it maintained, over two generations, for the purpose of general empirical experimentation with machines of all kinds, an "Inginary" of precisely the sort that Bacon envisioned in 1625. There, or in the neighborhood, were kept together a cadre of trained craftsmen under the supervision of sophisticated engineers and natural philosophers, and a physical plant with capacities for founding brass, forging iron, and soldering, turning, filing, and boring in metal, and for sawing, turning, and joining wooden models. Early on, abortive experiments with high-pressure steam pumps were conducted there by Worcester and Kalthoff, and later by Morland, and there, or in the neighborhood at least, Savery found the workmen with skills to make the first successful steam engine. All of these efforts were stages in the development of the paradigm of the steam pump implicitly stated by Drebbel and de Caus in the period 1604-1615, both of them for a time employees of the Crown (and in Drebbel's case an employee of the Ordnance Department).

THE ATMOSPHERIC PISTON ENGINE

Thomas Newcomen is universally and justly credited with having produced the first commercially successful steam engine. Savery's pump was not a success and only a few were installed: it was slow in operation, sometimes burst its seams at the higher pressures needed to drive water more than sixty feet or so above the receiver, and, because the receiver sucked water up from below, could not be constructed more than twenty feet or so above the source of water. In mines, this meant constructing and repairing a furnace, boiler, and receiver in cramped quarters and in

---

2: 325 is the source for Savery's statement that his experimental model was first demonstrated at "a Potter's House at *Lambeth*"; Cunningham 1897, p. 216, for the presence of potteries at Vauxhall around the turn of the century.

virtual darkness, and in deep mines in stages every seventy or eighty feet. Morland's pumps, operated by men or horses, probably were more reliable and just as efficient. But Newcomen's engine changed the balance, at the mines and reservoirs, in favor of steam. Newcomen's engine worked on a different principle. Instead of sucking water up into a receiver located somewhat less than the theoretical maximum of thirty-two feet above water level, it operated a traditional force pump (presumably in most cases of Morland's improved design) and thus could be adapted to many existing pumping systems. But its main feature was that the force was provided by atmospheric pressure rather than expanding steam. Instead of driving the water from the receiver by high-pressure steam (Savery's pumps sometimes were worked at pressures approaching ten atmospheres, or 147 pounds per square inch, if they did not burst beforehand), steam was used for the sole purpose of creating a vacuum in the cylinder. The cylinder was filled with steam from a boiler; cold water was then sprayed into the cylinder and the steam promptly condensed into water, leaving a partial vacuum in the cylinder; then atmospheric pressure, at 14.7 pounds per square inch, drove the ponderous piston into the empty cylinder, dragging with it a chain, and with the chain a traditional rocking beam on the other end of which hung a traditional force pump. Situated on the surface, close by the pit head, the Newcomen engine, although more complex to fabricate, was easier to operate, to maintain, and to repair; it was certainly safer; and it pumped more water. The first Newcomen engine was installed at the coal mines at Dudley Castle in 1712; half a century later, it was this engine that James Watt improved by adding a separate condenser, to create the modern generation of steam engines.[33]

[33] Like Savery, Newcomen left few papers and there is no full biography; for what information survives, see DNB "Thomas Newcomen,"

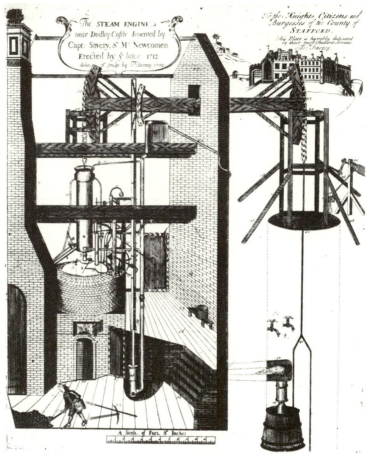

6. Thomas Newcomen's Atmospheric-Pressure Piston Engine (1712)

Newcomen's engine thus appeared as a new paradigm, sharing with the other line of engines only the general

Rolt 1963, and Jenkins 1971, Ch. 9, "Savery, Newcomen and the Early History of the Steam Engine." Early authorities, particularly Switzer 1724 and Triewald 1734, and most others after them, give Newcomen credit at the expense of Savery, and the interpretation offered here differs from the conventional wisdom on the matter.

notion of using steam to generate power. Newcomen's protagonists from an early date have claimed that the South Devon ironmonger and his partner, the plumber John Calley, originated the entire paradigm themselves, working for ten years independently of Savery (and of all the scientists who for half a century had been experimenting with the vacuum), to put the first production model in place at the coal mines near Dudley Castle in 1712. But some have been dubious of this alleged independence. John Robison, the Edinburgh physicist, claimed that Newcomen got the idea in correspondence with Robert Hooke of the Royal Society; unfortunately the letters allegedly held in the Society's collections have never been found. More recently, Joseph Needham drew attention, as many others have done, to the suggestion made in 1690 by the French scientific instrument-maker, Denis Papin, that the weight of the atmosphere be used to generate power by condensing steam within a cylinder and allowing air pressure to drive a piston into the resulting vacuum. This idea was a modification of Morland's, and of Papin's own mentor, Huygens', gun powder engine, which produced a partial vacuum in a cylinder by exploding gunpowder, thus allowing the piston to descend under the weight of the atmosphere. Papin published this suggestion of the atmospheric piston engine, with a diagram, in a Latin paper in the *Acta Eruditorum* of Leipzig in 1690, in French and Latin in 1695, and in French in Paris in 1698.[34] A brief abstract was published in the *Philosophical Transactions* of the British Royal Society in 1697:

> The fourth letter shows a method of draining mines, where you have not the conveniency of a near river to play the aforesaid engine; where having touched on the inconveniency of making a vacuum in the cylinder for this purpose with gunpowder, he proposes the alternately turning a small surface of water

[34] See *Acta Eruditorum*, August 1690: 410-414 and Tab. X Fig. 1.

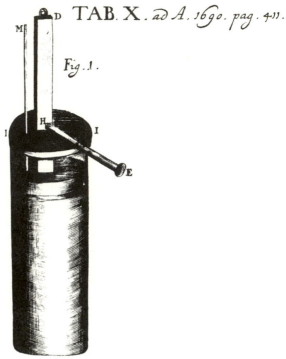

7. Denis Papin's Atmospheric-Pressure Piston Engine (1690)

into vapour, by fire applied to the bottom of the cylinder that contains it, which vapour forces up the plug in the cylinder to a considerable height, and which (as the vapour condenses as the water cools when taken from the fire) descends again by the air's pressure, and is applied to raise the water out of the mine.[35]

I cannot help but agree with Needham's opinion: "I find it almost impossible to believe that Newcomen did not know of Papin's steam cylinder."[36] By the same token,

[35] *Philosophical Transactions* 19 (1697): 155.
[36] Needham 1962-63.

56

one must add, it is even less possible to believe that Thomas Savery did not know of Papin's steam cylinder.

Denis Papin (1647-1712) was not an obscure continental mechanician unknown to English philosophers and engineers. After earning a medical degree in 1669, he moved to Paris and entered the laboratory of the physicist Christian Huygens, who was one of the original members of the Royal Academy of Paris, an institution founded in 1666 to carry out the program of Francis Bacon. Huygens, among other things, was interested in the improvement of the "air pump," a device used to produce a partial vacuum and thereby to investigate the properties of air by the study of the effect of airless space on fluids, plants, animals, and materials. Huygens set Papin to work improving the air pump, and the young man published a book on his improvement in 1674. In the same year Papin went to England. There he learned English, worked in Boyle's laboratory using his new air pump, translated Boyle's Latin into English for him, and made demonstrations of various devices before the Royal Society (including the pressure-cooker, which he invented). In 1680, on Boyle's recommendation, he was elected a member of the Royal Society.

Huygens, meanwhile, as early as 1674 had been experimenting with the idea of a piston-operated atmospheric engine that produced a vacuum by the explosion of small amounts of gunpowder in a cylinder (the hot expanding gases drove out the air, and as they cooled produced a partial vacuum). These experiments were known in England (perhaps Papin had brought news of them) but were hardly news there, for Morland had received a patent for a gunpowder pump as early as 1661. When Papin returned to the continent in 1688 he took up his experiments with the technology of producing a vacuum in cylinders and in that year published a paper on the gunpowder vacuum engine. As a devout Protestant, after the revocation of the edict of Nantes in 1685 he could not return to France, but

57

he received a professorship at Marburg. His time there, from 1688 to 1695, was his period of greatest production. He invented a centrifugal blower for ventilating mines, and continued to play with the gunpowder engine; it was directly out of this line of thought that the idea of the atmospheric steam engine developed and was published in 1690. From 1695 to 1707 Papin lived in Cassel and it was there, in 1705, that he received from his friend Leibniz a sketch of Savery's engine. He described a Savery-type engine of his own design and even constructed a steam powered boat (unfortunately destroyed by watermen on the Fulda, jealous of their rights). Papin moved back to England in 1708, hoping for help in developing his inventions; although he frequently participated in the Society's meetings, he received no aid. Savery, now a member of the Society, probably regarded him as a rival to be put down, and Newton, President of the Society, may have known of Papin's friendship with his enemy Leibniz. Papin died in poverty in 1712, the year Newcomen's first engine was put into use.[37]

It is hardly possible that Savery was totally unaware of Papin's earlier suggestions when he began to work on his steam pump. There is, furthermore, evidence that about 1701—after he had received a patent for the Lambeth engine—Savery was experimenting with a steam-operated piston engine in which high pressure raised a piston in a cylinder, and then gravity and air pressure drove it back down. Roger North—nephew of Samuel Morland's friend Francis North—in that year saw and sketched a Savery steam pump and at the same time described and sketched a piston engine, which however he "saw only in model." Although Henry Dickinson, Morland's partisan biogra-

[37] Although Papin has been memorialized in the standard French and English biographical and scientific dictionaries, the best source of technical, biographical, and bibliographical information about him is the collection of his correspondence with Leibniz and Huygens edited, with a biographical sketch, by Garland 1881.

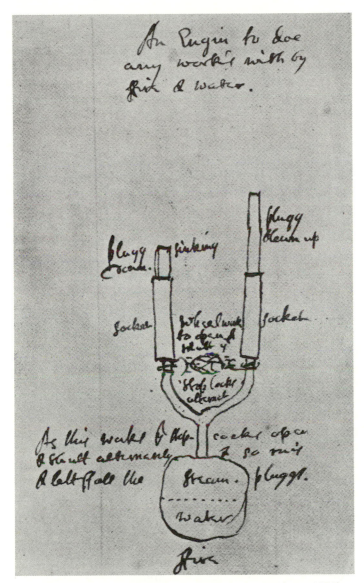

8. Thomas Savery's Steam-Pressure Piston Engine (circa 1700)

pher (and following him Rhys Jenkins) claim that North's visits to Vauxhall occurred some time between 1677 and 1682 (when Morland left for France), it is much simpler to suppose that North was simply recording two machines—a Savery steam pump and a model of a piston engine of Savery's design—that he saw on one visit to Savery's shop in 1701.

Savery's patent was a very general one, covering all engines that did work with the help of steam. Savery was explicitly interested in the use of his engine not merely to pump water but also to serve as a prime mover for ships and mill machinery. But the latter task required a reciprocating mechanical motion, like that of a piston, converted to rotary motion by a traditional crank, and the Savery engine had no piston. Papin's engine of 1690 did have a piston, and it is entirely likely that Savery was attempting to adapt his existing boiler and receiver system to the purpose of producing mechanical motion by using steam in the outflow pipe to drive a piston instead of water.

It is also not unlikely, at least, that Savery, his time occupied with the task of working out the bugs, and managing the business he had created, should turn over the task of developing the piston engine to a competent firm, along with his own ideas and other relevant material, including Papin's design. After all, his patent already covered the collateral paradigm also. Why he chose the ironmongering business of fellow Devonians Newcomen and Calley in Dartmouth is not suggested by the record, but Newcomen evidently accepted Savery's terms, for the Newcomen engine when it was perfected was sold by the Savery firm under the Savery patent. Savery never claimed that he was the inventor despite his patent coverage—a rare forbearance in an age of invention claim-jumpers—and barely a scrap of paper remains from Newcomen's hand. Perhaps Savery deserves more consideration than he is generally given in histories of technology, however.

Rather than a mere loser in the race for priorities, he might be seen as the man who, following Morland and Papin, developed the initial design for a practical piston engine and, when he turned the task of development over to another mechanician, generously refrained from claiming the resulting product as his own (contenting himself, no doubt, with a major share of the profits). Nor did Newcomen claim sole originality; the claims for Newcomen's priority were advanced—and challenged—by others.[38]

CONCLUSION

In concluding this long and somewhat tortuous review of the development of the steam engine in the seventeenth century, we have tried to be less concerned with individual priorities in the development of paradigms within a community of mechanicians, than with particular institutions within that community, and less concerned with the social attributes of innovators than with their forms of organization and support. Although the records available are scanty, we can gain at least the outlines of an image, like the shape of a building seen through fog (or, dare I say it, a cloud of steam?) of how research and development in the powers of steam and air was organized. In this perspective, we can say that the steam engine was developed by persons for the most part employed by or associated with a single institution, His Majesty's Office of Ordnance, over the course of nearly a century, particularly that part of the ordnance establishment that was with its cadre of Dutch artisans located in and around Vauxhall in Lambeth. At the end, of course, the inventors had a little help from their friends in the Royal Society of London, and, at the very end, from an ironmongery business in Dartmouth.

[38] There are several histories of the steam engine's development before Watt. In addition to the sources already cited, and the standard histories of technology, see Dickinson 1938 and Ferguson 1964.

II

Families of Iron

ONE OF THE CURIOUS FEATURES of Bacon's discourse on
the New Atlantis is the long discussion of marriage and
the family that immediately precedes the account of Salomon's House. To a twentieth-century reader it may even
seem a tiresome diversion to insert a long account of a
patriarchal feast and to praise the Atlanteans at length for
adhering to monogamy and to high standards of chastity,
for being, in the words of a Jewish merchant praising Bensalem, "the virgin of the world." But Bacon was a terse
essayist who would not introduce empty rhetoric into a
tightly argued analysis of an ideal system of society. Bacon
makes it plain that he is both following and in some points
diverging from Sir Thomas More's treatment of the family
in *Utopia*, which had been published little more than a
century earlier, and was well known to many of his read-
ers. More made the patriarchal extended family the economic and social basis of Utopian society. Utopian marriage was monogamous, premarital chastity and marital
fidelity were strictly enforced, the wife lived in her husband's household, along with her husband's patrilineage,
which generally included three generations and numbered
on the order of forty people. This group operated a kind
of rural agricultural commune but some of its members
resided temporarily in smaller urban family households.[1]

[1] Surtz and Hexter 1965.

63

Bacon's account is less extensive and his mode of presenting the system is to focus on a ceremony peculiar to the Atlanteans, the "feast of the family, as they call it." This feast not only celebrated the patriarchal values of transgenerational continuity and male dominance but—more importantly—led to an explicit contract between an extended family and the state:

It is granted to any man that shall live to see thirty persons descended of his body alive together, and all above three years old, to make this feast; which is done at the cost of the state. The Father of the Family, whom they call the *Tirsan,* two days before the feast, taketh to him three of such friends as he liketh to choose; and is assisted also by the governor of the city or place where the feast is celebrated; and all the persons of the family, of both sexes, are summoned to attend him. These two days the Tirsan sitteth in consultation concerning the good estate of the family. There, if there be any discord or suits between any of the family, they are compounded and appeased. There, if any of the family be distressed or decayed, order is taken for their relief and competent means to live. There, if any be subject to vice, or take ill courses, they are reproved and censured. So likewise direction is given touching marriages, and the courses of life which any of them should take, with divers other the like orders and advices. The governor assisteth, to the end to put in execution by his public authority the decrees and orders of the Tirsan, if they should be disobeyed; though that seldom needeth; such reverence and obedience they give to the order of nature. The Tirsan doth also then ever choose one man from amongst his sons, to live in house with him: who is called ever after the Son of the Vine. . . . The Tirsan cometh forth with all his generation or lineage, the males before him, and the females following him; and if there be a mother from whose body the whole lineage is descended, there is a traverse place in a loft above on the right hand of the chair, with a privy door, and a carved window of glass, leaded with gold and blue; where she sitteth, but is not seen. When the Tirsan is come forth, he sitteth down in the chair; and all the lineage place themselves against the wall, both at his back and upon the return of a half-pace, in the order of their years without difference of sex; and stand upon their feet. . . . Then the herald with three curtesies, or rather inclinations, cometh up as far as the half-pace; and there first taketh into his hand the scroll. This scroll is the King's Charter, containing gift of revenew, and many privileges, exemptions, and points of honor, granted to the Father of the Family; and is ever styled and directed, *To such an one our well-beloved friend and creditor*: which is a title proper only to this case. For they say the king is debtor to no man, but for propagation of his subjects.[2]

We need not pursue the details of the feast further, beyond remarking that Bacon went to great pains to describe the rich fabrics, the gold and jewels, and the religious fervor that marked the importance to the state of the extended family. Similar emphasis is given to the ritual accouterments of the "Father" of Salomon's House itself.

The significant point here is the "King's charter, containing gift of revenew, and many privileges, exemptions, and points of honor, granted to the Father of the Family," and the role of "the governor" in putting into effect the "decrees and orders of the Tirsan." What is happening at the ceremony is the public recognition of a self-governing kin group that is being given a royal charter. There were, of course, other chartered institutions in Bacon's England. Royal charters and patents of monopoly were granted by

[2] Bacon, *Works*, 1: 385-388.

the Crown (generally for an initial fee or royalty) to town corporations or companies of craftsmen (such as are noted briefly as existing in Bensalem in a reference to the "officers and principals of the companies of the city"), who might agree with an individual capitalist to be their exclusive agent under another royal patent; and some patents of monopoly were granted to private entrepreneurs, to partnerships, and even inventors. Bacon had been for years a member of the committee that advised the Crown on the granting of such patents. But recruitment into these organizations, which derived ultimately from the medieval guilds and now generally combined both the mercantile and manufacturing specialists of the same industry into one company, was not based on descent from a common ancestor (even if many tradesmen did follow their father's occupation).[3] There were many extended kin groups, including both the royal family and nobility and the landed gentry, who structurally might serve as the prototype for the Tirsan's descent group, but who required no special royal patent or charter to recognize their existence or to define their economic and political role. Such gentle families correspond in a general way to the anthropologist's model of a cognatic (non-unilineal) descent group, in this case one that owns an estate and permits any descendent of the founder to claim certain rights in the estate (or be assigned certain duties) by proving descent, through any combination of males and females, from the founder. Bacon significantly fails to mention purely male lineality; he simply says "descended of his body" and "all his generation or lineage" including both males and females. (Bacon, who had no sons but three daughters, all married, may have had a personal sympathy for a principle of personal continuity through either sex.) Bacon also uses the expression *stirp* (in referring to "a few

[3] See Unwin 1904 for an account of industrial organization in Bacon's time.

66

stirps of Jews" in Bensalem and, in the *Essays*, to "stirps of Nobility") which denotes explicitly a non-unilineal descent group or stock, particularly in the case of an estate entailed upon the descendants of an original owner.[4]

The strategic importance of Bacon's suggestion is that it provided a special institution for the execution of the state's economic plans (which were, presumably, framed with the advice of Salomon's House): the incorporated cognatic descent group governed by the Tirsan, claiming neither gentle birth nor title of nobility, but operating under a royal charter. Such families would not replace, as in More's *Utopia*, the other institutions below the state; they would simply exist and multiply alongside them, carrying out under patents of monopoly those new and special tasks that town corporations and conventional partnerships undertook only fitfully and unreliably. In Bacon's time, a hundred years before the limited liability stock company or corporation had appeared in England, his notion of incorporated families was not an unreasonable one.

EXTENDED FAMILIES IN THE IRON INDUSTRY

In one industry in Britain the corporate family and the newer manufacturing processes were coming into conjunction. That industry was iron. The smelting and forging of iron had been in late medieval times for the most part a rural affair, conducted by small operations of a dozen to twenty men, dependent on superficial local iron deposits for raw material, on forests and coppices to provide the charcoal to fuel the bloomery furnaces and forges, and on the smaller streams to turn the water wheels that worked the bellows and heavy machinery. The blast furnace was introduced from Germany in the early sixteenth century and led to an expansion of the industry. In the latter part

[4] Bacon, *Works*, 1: 390; 2 part 2: 121.

67

9. Diagram of a Blast Furnace

of the sixteenth and the first half of the seventeenth century, the largest concentration of blast furnaces was located in the Weald, a forested area south of London in the
counties of Kent and Sussex. As the industry gradually
grew in size, furnaces became more numerous in the Midlands, Wales, and the north of England. The furnaces produced some pig for conversion into wrought iron, but much
of the product was directly cast, or founded, on the spot,
particularly into the iron cannon required in large numbers by the British navy and by British merchant vessels;
when permitted, such cannon were also exported to France
and Holland.[5] These cannon were, in Bacon's time, cast
hollow so that the bore had only to be cleaned out, and
perhaps roughly drilled or filed, before sending on to the

[5] Cipolla 1965.

68

Tower of London to be proven. Furnaces also produced a variety of cast-iron domestic products, such as fire-backs, ovens, kettles, and pans. Forges producing wrought-iron objects were sometimes part of the same business, but they depended in large part on imported Swedish and Russian iron, for England mined enough iron to supply less than half her own needs. Forges reheated pigs and turned them into wrought iron, to be rolled into sheets in rolling mills and cut into strips in slitting mills. The large, combined enterprise that could do all of these operations was more efficient than the smaller twelve-man business that mined, cut wood, made charcoal, and smelted iron at a bloomery. But the bigger furnaces also required expensive capital investment, careful bookkeeping, and elaborate industrial management of, in some cases, hundreds of workers. Furthermore, both economy and control of the market favored the larger business that could consolidate a number of works under its administration.[6]

In Elizabethan times, one institution that undertook this task was the partnership under a royal patent of monopoly. Bacon, it may be recalled, had himself been a partner in a group that operated the famous ironworks at Tintern in Monmouthshire, and a brief review of this enterprise may be helpful in clarifying the contrast between partnership and family principles in the iron industry of the following centuries. With the support of Lord Burghley—one of Bacon's predecessors as Chancellor—about 1565 a patent of monopoly was granted to one William Humfrey, Assay Master of the Royal Mint in the Tower, for the erection of the first English wireworks operated by water power. The works at Tintern were originally intended to draw brass wire, but brass being then new to England, the works turned to the manufacture of iron wire. In 1568 a partnership named "Society of the Mineral

[6] Standard histories of iron and steel manufacture in Great Britain are Schubert 1957 and Ashton 1963. See also Hodgen 1952, Flinn 1958, Hunt 1954, Jenkins 1925-26.

69

and Battery Works" (of which Bacon was a member) took over Humfrey's patent and proceeded to develop a successful and lucrative wireworks. The wire was used all over England to make all sorts of goods—mousetraps, wool cards, needles, bird cages, and whatnot. This monopoly was maintained against infringement for at least fifty years. In addition, the Tintern forges introduced the method of making iron plate by the use of a water-powered hammer (the "battery").[7]

A contrasting case—certainly known to Bacon, both from his administrative position and because of the personal interest he had in the iron trades—was the endowment of a family of iron: the Browne family of the Kentian Weald. The Office of Ordnance from the middle of the sixteenth century had contracted for its cast-iron guns with the holder of a royal patent of monopoly. The first holder of such a patent was Richard Hogge, who (it was said) cast the first iron gun in 1543 and thereafter served as Her Majesty's "Gunstone Maker for Life . . . onlye cast for the Tower." Hogge was succeeded in 1589 by Thomas Browne, who was also appointed royal gunstone maker for life. In 1614, Thomas was succeeded in this lifetime monopoly by his son John Browne, who was able to retain his position as sole maker of naval ordnance even when Parliament replaced the King. John's sons continued as gunfounders, and one of *their* sons was renewed as the King's gunfounder as late as 1681.[8] The Browne family, in summary, had been endowed with a virtual monopoly of naval ordnance extending over four generations and at least one hundred years. The Brownes may well have been in Bacon's mind when he was describing the perquisites of the Tirsans of Bensalem. Another family in iron at an early date, of course, was the Dudleys of Worcestershire, who as we saw in the last chapter anticipated the Darbys

[7] Schubert 1957, 292-304; Rees 1968.
[8] Victoria County Histories of England, Kent, 3: 384-391; Straker 1969; Ffoulkes 1937; Tomlinson, 1976.

of Shropshire in smelting iron with coal as early as 1619, when they received a royal patent for the process. (From Elizabeth's reign onward, the crown encouraged the substitution of coal for wood in all possible industrial processes.) The Earl of Dudley and his line may well also have occurred to Bacon when he was reflecting on the economic role of the family; but we shall defer discussing them further. They will reappear in the next chapter, among the innovators in the coal trade.

From Bacon's time on, the role of extended families in English industry constantly increased. To be sure, during the seventeenth century, the loose standards of sexual morality in court circles might have suggested to some that the institution of marriage was in peril, and such fears were loudly voiced at the time. But neither Puritan nor Royalist was in any real doubt about the economic sacredness of marriage and descent, to which were ordinarily attached important alliances and estates in land, in business investment, and even in such intangibles as secret recipes (like Drebbel's scarlet dye, passed on by word of mouth to his sons-in-law). Bacon in his *Essays* might speak rather lightly of the English family as a domestic institution, alleging that "the Noblest workes, and foundations, have proceeded from *Childlesse Men*"; and he regarded a wife, no less than children, as "Impediments, to great Enterprises." But then, he also regarded the English system of patrilineal primogeniture poorly in comparison with the Italian:

The Italians make little difference between children and nephews or near kinsfolks; but so they may be of the lump, they care not though they pass not through their own body. And, to say truth, in nature it is much a like matter; insomuch that we see a nephew sometimes resembleth an uncle or kinsman more than his own parent.[9]

9 Bacon, *Works*, 2 part 2: 100.

71

By the 1690s, a large proportion of the English iron production was controlled by great families, some of which—like the Brownes—had already been in the iron business for generations. In the Midlands, around the turn of the century, no less than fifty furnaces, forges, and slitting mills were controlled by members of the Foley family, including the wireworks and foundries at Tintern in which Bacon had earlier held an interest.[10] In Wales, the Guests controlled the great Dowlais ironworks.[11] Centered in Stourbridge, in the seventeenth century the Crowleys developed a vast nationwide organization, with the largest ironworks in Europe located near Newcastle, offices in London, and a warehouse and transportation network all over the country; for fifty-four years, at the height of its prosperity in the mid-eighteenth century, the business was directed by a woman, the widow of John Crowley, who had died at the age of thirty-eight. The Crowleys will be mentioned later in connection with the remarkably advanced system of industrial organization that Theodosia Crowley administered.[12] Another well-known iron family was the Crawshays, who developed the ironworks at Cyfartha, not far from Dowlais, in South Wales. The Crawshays were self-conscious about the similarity of their line, the "Crawshay Dynasty," to a noble family, and referred to kinsmen by such descriptive titles as "His Majesty" and "Vice-Roy." Like the others, the Crawshays' control of their industrial empire survived for more than a hundred years.[13]

There were many other industrial families in this era, both in iron and in other fields; but rather than attempt a census, let us try to describe their mode of organization. What were the principles by which these great industrial families were constructed, and how did they relate to the

[10] Johnson 1951.
[11] Wilkins 1903.
[12] Flinn 1962.
[13] Addis 1957.

72

organization of business? The Crowleys and the Crawshays will provide perhaps the clearest example. We are
fortunate in being able to depend on an excellent study
by the historian Michael W. Flinn—*Men of Iron: the
Crowleys in the Early Iron Industry.* Flinn provides a good
deal of the detail necessary to describe how the Crowleys
were organized; and he also gives a genealogy showing
five generations of Crowley marriage and descent, which
when matched against the history of partnership rights
and administrative duties reveals to some extent the connection between business, marriage, and descent. The
Crawshays have been memorialized in an equally detailed
account—*The Crawshay Dynasty*—by John Addis.

The first Tirsan of the Crowleys, Ambrose I (c 1605-
1680) was a nailmaker in a small village near Stourbridge;
he had at least five sons and four daughters, and died
possessed of a six-room house, workshop, and a barn. One
of the sons, Ambrose II (1635-1721), moved to Stourbridge,
joined the Society of Friends, prospered there as an ironmonger (i.e., a wholesaler of iron), and produced fourteen
children, including Ambrose III, the only child by his first
wife. Ambrose II expanded beyond his ironmongering
business, leasing at least two forges in the neighborhood,
where he made rod-iron for the nail trade and produced
steel in a cementation furnace. He was also involved in
enterprises in Wales and urban waterworks in Devon. Ambrose II's iron enterprises in Stourbridge became inter-
nationally famous and he was remembered in later years
as having helped mightily to build up the iron industry
in the Stourbridge area.

With such extensive undertakings to manage, Ambrose
II turned to his family for assistance. About 1680 his eldest
son, Ambrose III, became his second-in-command. Two
of his daughters married two brothers, both of whom participated in managing the family business. (A third daughter, Judith, was unsuccessfully courted by James Logan,
William Penn's secretary in Pennsylvania, and she even-

73

tually married a relative of Samuel Johnson, who visited the Crowleys in Stourbridge and wrote verses for Sampson Lloyd's daughter—another iron family.) The three eldest sons of the second marriage left Stourbridge and lived independently: one went to sea; one lived in London as apprentice to a draper; the third moved westward to set up an iron furnace in South Wales. The three youngest sons, however, were employed in the family business, and when it was formally divided into a partnership in 1710, the parts were controlled by different members: Benjamin managed the ironworks in Wales, John a mill for grinding dye-woods, and James the steel trade. The whole enterprise was coordinated and directed from London by Ambrose III and Sampson Lloyd.

After the death of Ambrose II, Ambrose III took charge. First establishing himself in a successful ironmongering business of his own in London, he expanded into the supply of sheathing nails to the British navy, and that in turn led by degrees to the putting together of the large system of iron manufacture and supply that was to earn him fame and a knighthood. In the Valley of the Derwent, in Durham, a few miles west of Newcastle, from 1697 on, Ambrose III built a rolling and slitting mill, two steel furnaces, a plating forge, warehouses, factories, offices, and workmen's houses, and set up the extensive organization of transportation and retailing that centered in his London office.

Ambrose III left only one surviving son and four daughters when he died at the age of fifty-five in 1713 (five children died in infancy), and the daughters had married prominent men not involved in the iron business. His son, John I (1689-1728), however, built up the great Crowley system even further; and when he died in 1728 at the age of thirty-eight, his widow Theodosia, later assisted by her two sons, Ambrose IV and John II, continued to manage the enterprise. The business was still intact in the first decade of the nineteenth century.[14]

The story of the Crawshay dynasty begins in London in 1772, when Richard Crawshay (1739-1810), a successful wholesaler of iron products, took as his partner one Robert Moser; next year the partner married Richard Crawshay's sister Elizabeth (1747-1824). Three years later a Roger Moser (probably Robert's brother) married Sarah, another of Richard Crawshay's sisters. Robert Moser died in 1785 but his son Robert eventually took an important position in the business. In 1790 the widow married Robert Thompson who was promptly taken into the firm as a partner. Thompson left the firm in 1798 to take over an ironworks at Tintern, leaving Richard Crawshay and his son William I (1764-1834) as the sole partners. In the meantime the firm had been prospering with the help of other, temporary partners, and had acquired a major interest in a large Welsh gun foundry at Cyfartha; when the original owner of that works died, Richard and William Crawshay became sole lessees.

Other kinfolk of William I's generation were associated with the Crawshay enterprises at Cyfartha. In 1774 Richard's third sister, Susannah (1747-1812), had married John Bailey, and their two sons came to work for their uncle, one as a partner and the other as a trusted employee. William I's sister Charlotte married Benjamin Hall, a barrister who represented the district in Parliament for a time and who took an interest in the iron trade; when Richard died, he left Hall a share in the works that made him an active partner. William I's cousin Elizabeth, the daughter of Elizabeth Crawshay and Robert Moser, married one George Pulling, whose sons Robert and Wellington became partners in the 1830s. The data could be amplified, for the Crawshay family maintained control of the Cy-

[14] In addition to Flinn 1962, see Flinn 1957 and Young 1923-24.

fartha Ironworks until 1902, when it came under the control of the equally ancient Dowlais firm, which was headed by another powerful family of iron, the Guests.[15] There are obvious similarities in the kinship practices of the Crowley and Crawshay families: the marriage of a pair of brothers to a pair of sisters; the granting of partnership, or at least employment, to the husbands of Crowley and Crawshay women; the inheritance of the position of *head* of the firm in the patrilineal line; the recognition of claims to participation in the firm by cognatic descendants (i.e., the descendants of Crowley and Crawshay women); the maintenance of effective administrative control of the business over a number of generations (in the Crawshay case, no less than five). The "family" of relevance to the firm, therefore, is not just the lineage that bears the name of, and is directly descended in the male line from, the original male founder of the firm, but also includes a ramifying network of men and women connected by marriage and descent through one or more females from the original founder. It is a non-unilineal descent group, all of whose members and their spouses (and even their spouse's siblings) have some claim to interest in an estate that bears the name of a business firm founded by the titular ancestor; only those who can validate this claim by work, residence, or financial contribution, however, are recognizable as part of "the family firm" at any one time.[16]

The Crowleys and the Crawshays are similar in another respect. In order to maintain a well disciplined and stable work force, they both established, or tried to establish, a paternalistic relationship to the communities in which their works were located. In the case of the Crowleys at Winlaton in the early eighteenth century, there was an

[15] Addis 1957.

[16] For a useful introduction to the concept of non-unilineal kin group and various forms that it may take, see Fox 1967, particularly pp. 146-174.

elaborate set of written administrative laws, with a system of courts and appeals, that governed factory discipline and discipline over home-workers; the regime included an explicit welfare program financed by joint weekly contributions by workers and the employer. Schools were set up for the workmen's children, the chaplain was assigned to function as an arbitrator as well as to provide religious services; a full-time salaried physician-and-surgeon was employed to diagnose and prescribe medicines and provide medical services for employees without charge. The Crawshays, nearly a century later at Cyfartha, played a similar role in their community of several hundred worker families. Richard Crawshay, the so-called "King of Myrthir," was a devoted churchman who built and endowed a chapel for the community of Cyfartha and induced Robert Raikes to establish one of his Sunday schools for workers' children. He refused to exploit his workers by the truck system of paying wages in goods instead of money and (in defiance of the Corn Law) attempted to sell his workers cheap American flour.

The outline that emerges, even from the incomplete records available, is of a type of kinship structure that resulted from the practice of raising capital and recruiting managers by forming a partnership and at the same time ensuring the loyalty of the partner by marriage to one's sister or daughter. Over the course of several generations, a complex web of affiliations between firm and family would be created that made constantly available a supply of active participants in the business who were motivated not merely to make money for themselves but for their descendants. While ideal solidarity can hardly have been expected—and there were in fact bitter disputes over business policy in the Crawshay family—the successful maintenance of two firms for a hundred years in the Crowley case and one hundred and twenty-five for the Crawshays suggests that the transgenerational character of the family concept was applied to the firm itself. Furthermore, the interest of the firm in the welfare and stability of the workers' community had an almost feudal, manorial quality.

This is not to suggest that this structure was invented *de novo* by the seventeenth-century and eighteenth-century capitalists. The relationship between descent group and landed estate was the crux of the Scottish clan system, which allowed an individual to claim residence on any one of several clan grounds on the basis of reckoning through either males or females; and similar practices may have been common in England as well, but without the advantage of Scottish terminology and heraldry. Large mercantile and banking houses, like those of the Fuggers and Rothschilds on the continent, had long relied for solidarity on the ties of marriage and descent. Firms founded on such a basis may well have had a real competitive advantage and simply absorbed more opportunistic partnerships. So it is probable that the extended family mode of business organization was an old and well-known option for English entrepreneurs, given new currency now in the opening years of the Industrial Revolution. It would have a peculiar affinity for industries such as iron, where massive works and large, permanent communities, living on extensive forest, coal, or mineral lands, easily established a semi-manorial way of life—or, in the colonies, a "plantation."

The Darbys and The Coalbrookdale Company

We have not so far broached the topic of the relation between the family mode of industrial organization and the support of technological innovation. Both the Crowleys and the Crawshays, but especially the latter, would be interesting to consider from this standpoint. We will, however, turn to another great industrial family, the Darbys, whose name is associated with some of the major technological innovations in the iron industry during the eighteenth century. The founder of the firm, Abraham Darby I (1678-1717), established a brass foundry in Bristol in 1699 that quickly turned to making iron castings. In 1707 Darby moved to Coalbrookdale in Shropshire, along the banks of the Severn River. The ironworks and associated coal and iron mines in Coalbrookdale and several surrounding villages were owned and managed by five successive generations of Darbys; the last managers bearing the Darby name departed in 1851 (but only to take posts as managers at other ironworks); and the last Darby shares in the business were relinquished in 1922. Let us confine our attention to the one hundred and fifty years of combined Darby management and ownership, however, and examine the kinship structure of this Quaker family firm more closely (see Fig. 10 for a genealogical/organizational chart to which reference will be made repeatedly). Even with the limited amount of genealogical and organizational data available, it will be quickly seen that the structure is complex, for we are looking at the relationship between kinship status and business role in a group of several hundred people (of whom, however, we have available the names of only about fifty). There is, to begin with, a set of individuals who may be termed the Darby connection, who are related by descent from, or marriage to descendants of, the John Darby of Worcestershire who was the father of Abraham Darby I, the founder of the firm. The Darby connection thus abstractly defined (and revealed in published genealogies, which do not necessarily correspond to the mental charts of the Darbys themselves) is at its core a non-unilineal descent group that includes not only persons who bear the name Darby but also descendants of Darby women; furthermore, it includes the wives of Darby-connection men (who may or may not be descendants of John Darby), and also the husbands of Darby-connection women (who also may or may not be descendants of John Darby) and even their relatives by de-

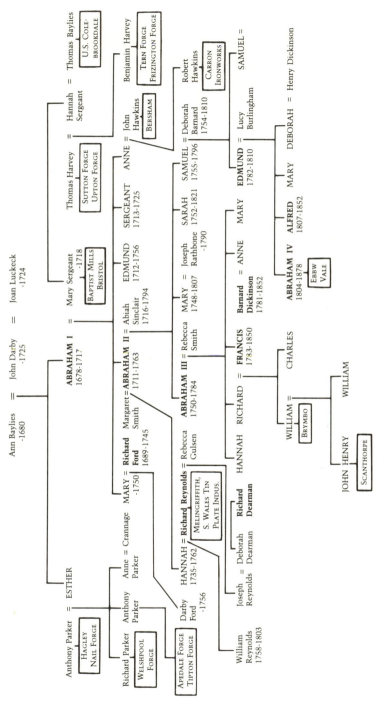

10. A Partial Genealogy of the Darby Connection

MANAGERS OF COALBROOKDALE CO.

ABRAHAM I	1699-1717	EDMUND	1803-1810
RICHARD FORD	1717-1745	BARNARD DICKINSON	1810-1827
ABRAHAM II	1738-1763	FRANCIS	1810-1850
RICHARD REYNOLDS	1763-1768	ABRAHAM IV	1827-1849
ABRAHAM III	1768-1789	ALFRED	1827-1851
RICHARD DEARMAN	1792-1803		

scent and marriage as well. Next, there is a set of individuals who are members of a firm that may be termed the Coalbrookdale Company. By members of the firm I mean to include, first of all, the several partners who hold shares in the business and who meet as a board of directors, casting votes on company policy and personnel matters according to the number of their shares; and second, the senior administrators of the firm, hired by the board to direct the business—managers of particular works or divisions (furnaces, foundries, mines, transportation), and outside agents selling wares and buying supplies. The partners were not all members of the Darby connection, at the beginning, nor were all of the managers and agents. But the board over the years came increasingly under the control of the Darby connection until all the shares were in the hands of members, and most of the managers and many of the agents were Darby-connection people too.

The interesting question is, what were the criteria by which Darby-connection persons on the board selected other Darby-connection members as members of the board or as managers and agents? How did a Darby define a Darby connection? How did he define his own kindred (including relatives not necessarily members of the Darby connection)? One would like to have access to unpublished correspondence, diaries, and meeting minutes; but even without examining these primary sources, some inferences can be drawn from available materials. First of all, membership on the board was determined by ownership of shares; and inasmuch as shares could be inherited, the provisions of a will, or the lack of one, determined the composition of the board after the death of a board member. Two principles appear: first, both males and females inherited shares, the female owners including widows and daughters; and second, not all children were given shares in the disposition of an estate. Another means of acquiring membership on the board was by marriage. Again, two principles appear: a non-Darby man could "occupy" his Darby wife's shares; and a share-owning Darby man's widow inherited part of his share (part went to his children). Shares could be sold in order to raise capital for the firm, and the preference of course was to sell shares to a member of the Darby connection, whether or not already a shareholder. A member of the firm who sold his share might also do so in order to take the capital for an independent venture.

The result of these, and perhaps other, principles of decision was the development over the generations of an ownership-management-and-family group that, in effect, defined at least minimally what the Darbys considered the Darby connection to be. It was, first of all, not a patrilineal descent group, and inheritance of shares was not governed by a principle of primogeniture. Although males in fact occupied all the managerial and agent positions, women owned shares in their own right and passed them on to their children, whether their name was Darby or not. The readiness to reckon through female connections may have been facilitated by the Darby's Quaker ethic, which allowed women roles of leadership in religious meetings; such an ethic would also have been compatible with the presence of women as shareholders. But this in itself was hardly an astonishing liberalism in a country that (in contrast to France) was perfectly willing to have a woman wear the crown and to reckon dynastic claims through females as well as males. The Crawshays and the Crowleys were not Quakers yet they seem to have proceeded along similar lines. More importantly, this way of defining family presumably had the advantage of making available for membership on the board a wider range of capital, talent, and commitment than a strict adherence to a patrilineal succession could have done.[17]

[17] The principal source of information on the Darby connection is Raistrick 1954.

83

Some light is shed on Darby family attitudes in their literary productions: diaries, letters, and published works. It is clear that, although Darby women did not participate in the day-to-day management of the business, they were powerful influences in determining the intellectual and moral atmosphere of the Darby-connection families around Coalbrookdale. The Darbys were evangelical Quakers and Abraham Darby II's second wife Abiah (a youngish widow of 29 at her marriage in 1745 to the head of the Coalbrookdale firm) took advantage of the traditional readiness of Friends to accord equality to women in religious matters. She was an itinerant evangelist who traveled thousands of miles on horseback, sometimes alone, to preach the word of God to the ungodly at military garrisons, town halls, chapels, and public markets. On her travels she met and dined with the wealthy and powerful, and she carried on an extensive correspondence; her house was a place where prominent travelers in the neighborhood were apt to stay. She read widely and took an interest in astronomy. She published religious tracts, including an "Expostulatory Address" opposing horse-racing, cockfighting, gaming, plays, dancing, musical entertainments, and other "vain diversions." She was concerned that children receive a Christian education, publishing her own textbook and eventually supporting the establishment of a Robert Raikes Sunday School at Coalbrookdale. She and the rest of the connection cooperated with the local Anglican "priest" to keep sin out of Coalbrookdale.[18]

Abiah's step-daughter Hannah married another Quaker ironmaster and tin-plate magnate, Richard Reynolds "the philanthropist," who thereby became a shareholder in the Coalbrookdale Company. From the edition of his family correspondence edited and published by his grand-daughter, we learn more about the attitudes internal to the Darby connection. Although from a philanthropic stand-

[18] Darby 1913.

84

point he considered all mankind as "children of one family," he admitted a selectivity in personal favor. "Our relative connection," was the term he used for the network of his own kin (his "kindred") and he candidly stated that, among those equally near in consanguinity, he nevertheless felt "a diversity in the mode or degree of our affection." Women he considered to be "meek" and "softer" by nature than men; he regarded himself, despite the profession of Quaker principles of friendship and pacifism, as prone to anger and severity. He displayed much of the classic Protestant ethic, demanding punctuality of himself, obedience from servants, and education for all. Over the fireplace stood the inscription "A Place for Everything and Everything in its Place."

Richard Reynolds's attitudes on the property relationships between husband and wife were, however, strictly in line with a principle of the concentration of capital in the Darby connection. When he married Hannah Darby in 1857, he moved from his house in Bristol to Ketley near Coalbrookdale, where he owned a furnace, and made a legal arrangement that settled his wife's shares in the Coalbrookdale Company upon her children in the event of her death. (Under English common law, her property was irrevocably merged with his unless such a settlement was made.) She did in fact die five years later, leaving two children, William and Hannah, who inherited their mother's property. When Abraham Darby II died the following year, Reynolds was made superintendent of the firm although he now held no shares. Later on, in advising his recently married nephew on the financial obligations of a husband, he discussed the moral importance of a related issue: making a settlement upon the wife of the property she brought to the marriage, so that, in the event of *his* death, her estate would be exempt from the claims of her husband's creditors. During their mutual life, he explained, he "occupied" her fortune and had administrative control over her property. But the settlement effectively prevented him from disposing of her property to others. Inasmuch as in his own case this property consisted in large part of shares in the ironworks, the settlement had the effect of keeping ownership and control of the company in the hands of the Darby connection.[19]

It would be fascinating to examine in more detail various other aspects of the internal dynamics of the Darby connection, and other extended family units in the iron industry. One would like to know how marriages were arranged, how wills were composed, how shares in new enterprises were assigned. It would also be interesting to compare the iron family "connection" with other varieties of the non-unilineal descent group, such as the Scottish clan and the cognatic lineages of Oceania, that also had the function of relating an estate to a non-unilinear descent group. But these topics would take us far afield from our purpose here, which is to consider these organizations as settings in which innovation was nourished. In view of the fact that the iron industry in England in the seventeenth, eighteenth, and nineteenth centuries seems to have been dominated by large family firms, we must ask, was there something about the familistic character of the enterprises that gave them an advantage by encouraging innovation in a highly competitive and rapidly evolving industry?

There is no question that the Coalbrookdale Company contributed importantly to technological progress. Its most celebrated contribution—the smelting of iron with coal—made its appearance at the very beginning of the company's history, and the firm continued to develop the technique and its application for another fifty to seventy-five years. About 1708 Abraham Darby I developed a process for successfully smelting iron ore with coke. Coke is bituminous coal from which sulfur and hydrocarbons have been driven off in a process analogous to making charcoal

[19] Reynolds family letters are published in Rathbone 1852.

from wood. Actually, efforts had been under way in England for at least a century for the making of iron with coal, which was already extensively used for other industrial heating purposes, but they were uniformly unsuccessful for several reasons, one of which was that the coal was not coked.[20] Raw coal's impurities were absorbed into the iron, and it was in consequence too brittle for casting and difficult to work into forged iron even when re-melted. There were at least four major advantages in the coke-smelting process: most of the impurities had been driven off, leaving nearly pure carbon; coke is more resistant to crushing than charcoal and thus the furnace could be larger and output increased; the temperature of the coke furnace was higher, and a less viscous liquid metal could be produced, which was much easier to cast; and, at least in some areas, coal was cheaper than wood because English forests were being used up by the multiple demands placed upon them for domestic heating, charcoal making, and naval timber. The long-run implications were profound, but Abraham Darby was not extensively imitated at the time, even though his pot-and-kettle business prospered, and he did not bother to secure a patent (in contrast to his seventeenth-century predecessors, several of whom took out patents for making iron with sea-coal or pit-coal).

Abraham Darby's success is just the beginning of the coke-and-iron story, however. The first diffusion of the new process happened in 1721, when the Lloyd family adopted it at their works in Bersham, adjacent to Coalbrookdale. In 1726 these works were taken over by John Hawkins—Abraham Darby's son-in-law—with financial help from the Coalbrookdale Company; and he in turn was succeeded by Isaac Wilkinson and his sons, who manufactured cannon (the pacifist Darbys would have nothing to do with the Office of Ordnance) and were later to be

[20] Celebrated earlier failures are described in Sturtevant 1612, Rovinson 1613, and Dudley 1665.

87

the principal makers of the Boulton and Watt steam engine. In the 1750s knowledge of the process was carried by Guest of Broseley—another nearby foundry—to Dowlais in South Wales. In 1760, when the great ironworks at Carron were set up in Scotland, it was John Hawkins's son from Bersham who was hired to manage the coke smelting.

Although coke iron produced by the Darby process made satisfactory pots and kettles, stove parts, and iron grates, it was still impregnated with too much sulfur and other impurities to yield satisfactory forged iron. The production of wrought iron thus remained dependent on charcoal. In view of the vast market for coke iron if it were suitable for refining, much thought was given to the problem at Coalbrookdale. By 1749, according to his widow Abiah, Abraham Darby II had developed a technique for making pit-coal pigs fine enough to match charcoal pigs at the forges. But the forges themselves still used charcoal, an increasingly expensive fuel. In order to solve this problem, "sometime" before 1766 two foundry-masters employed by the Coalbrookdale Company (the wife of one of them was a daughter of Abraham Darby I's sister), Thomas and George Cranage, suggested the use of a reverberatory furnace in which pigs could be melted in a hearth over a coke or even a raw coal fire where, separated from the coal itself, impurities could not be added to the metal. Richard Reynolds, now manager of the firm, had already come reluctantly to the opinion that charcoal was necessary to the process, but the Cranages argued that all that was needed was heat, and they asked leave to "make a trial." Reynolds, who seems to have been a remarkably honest and even generous man, provided a succinct account in a letter to an associate in Bristol:

I consented, but, I confess, without any great expectation of their success; and so the matter rested for some weeks, when it happening that some repairs had

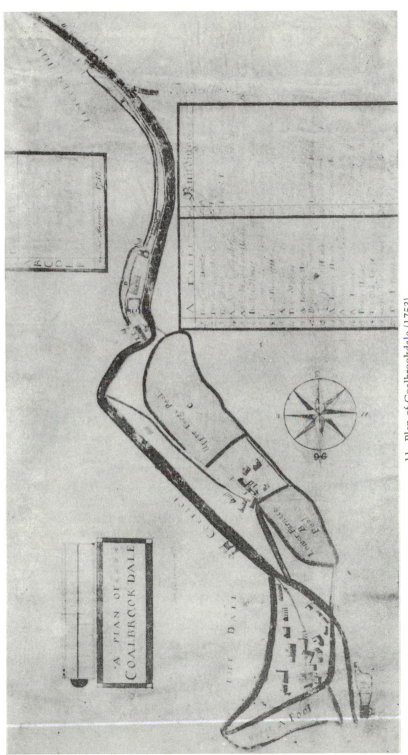

11. Plan of Coalbrookdale (1753)

12. View of the Upper Works, Coalbrookdale (1758)

to be done at Bridgnorth, Thomas came up to the Dale, and, with his brother, made a trial in Thos. Tilly's air-furnace with such success as I thought would justify the erection of a small air-furnace at the Forge for more perfectly ascertaining the merit of the invention. This was accordingly done, and trial of it has been made this week, and the success has surpassed the most sanguine expectations. The iron put into the furnace was old Bushes, which thou knowest are always made of hard iron, and the iron drawn out is the toughest I ever saw. A bar 1¼ inch square, when broke, appears to have very little cold-short in it. I look upon it as one of the most important discoveries ever made, and take the liberty of recommending thee and earnestly requesting thou wouldst take out a patent for it immediately. The specification of the invention will be comprised in a few words, as it will only set forth that a reverberatory furnace being built of a proper construction, the pig or cast iron is put into it, and without the addition of anything else than common raw pit coal, is converted into good malleable iron, and being taken red-hot from the reverberatory furnace to the forge hammer, is drawn into bars of various shapes and sizes, according to the will of the workmen.[21]

The patent was taken out in the name of the Cranage brothers. The Cranage process, involving the heating of coke iron with raw coal in a reverberatory furnace and stirring or "puddling" it with iron rakes was extensively adopted and improved over the next twenty years. It was finally perfected by Henry Cort, who in 1785 substituted rollers for the forge hammer in beating out the slug, and completed the basic process of manufacturing wrought iron that still is in use today.

The contribution of the Coalbrookdale Company to the making of iron may be stated succinctly: it was in Coalbrookdale works that the entire transition from wood to coke as the fuel for iron-making was invented and successfully introduced, in a period of sixty years, under the management of three members of the Darby connection: Abraham Darby I, Abraham Darby II, and Richard Reynolds. Each contributed, or supported, an essential step: (1) Abraham I first successfully used coke for smelting iron for casting; (2) Abraham II introduced improvements that made coke-iron pigs usable for refining into wrought iron in charcoal-fired forges; (3) Richard Reynolds provided two of his employees with the equipment for the development of the reverberatory furnace and puddling process that used raw coal for making wrought iron from coke-iron pigs. This cumulative accomplishment resulted directly in a vastly increased demand for coal and must be ranked with the development of the steam engine as one of the major technological transformations of the Industrial Revolution.

[21] Quoted in Raistrick 1954, 86-87.

13. Smelting House (1788)

But there were other lines of innovation to which the Coalbrookdale Company contributed. Shortly after the introduction of the Newcomen engine in 1712, the Coalbrookdale firm, under the direction of Richard Ford, another of the connections by marriage, began to improve upon it, particularly in substituting cast-iron cylinders for Newcomen's original copper ones, and improving boring techniques so as to have a reasonably smooth and true cylindrical shape. The firm for a time had a virtual monopoly on the provision of all of the parts of which the steam engines were made, together with the pumps and pipes. Even after the development of the Boulton and Watt engines, they continued to play an important part in the steam engine business, although secondary to that of the Wilkinsons, whose greatly improved—and patented—boring mill earned them the bulk of Boulton and Watt's contracts. The Watt engines quickly superseded the Newcomen design, with Wilkinson manufacturing the cylinders but the Coalbrookdale Company continuing to make pump and pipe parts. Coalbrookdale, with the aid of Hornblower and Trevithick, also experimented with the possibilities of a two-cylinder high-pressure modification of the Newcomen engine. In 1802 the company was developing, under the direction of Richard Reynolds's son William, a steam-powered locomotive for use on the company's extensive system of railroads. They also produced an experimental steam barge designed by Trevithick.

The Coalbrookdale Company was an innovator in still other ways. Its method of casting iron in molds of sand, instead of loam or clay, is still in use in modern foundries; the Company introduced cast-iron rails for railways; and, of course, in 1776 the company began construction of the famous iron bridge, first in the Western world and still standing across the Severn (and tolls from which were still being collected by members of the Darby connection as late as 1950). But enough has been said to make it plain that the Coalbrookdale Company, under the management

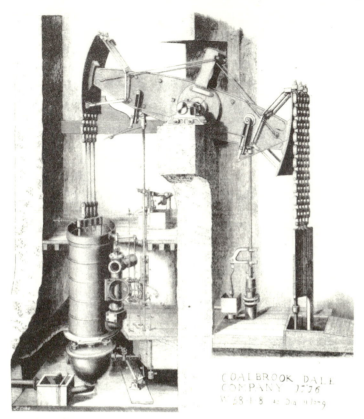

14. Newcomen Engine at Coalbrookdale (1776)

of the Darby connection, for at least a century continuously experimented with improvements in iron foundry processes and products, and invented and introduced fundamentally important innovations whose technical and social consequences have been vast.[22]

[22] The contributions of the Darbys of Coalbrookdale to the iron industry have been described and evaluated in a number of places. In addition to Randall 1880, Ashton 1924, and Raistrick 1954, see also Ashton 1926, Hall 1926, Jenkins 1923-24, Johnson 1960, Court 1938, and Trinder 1973.

15. Ironbridge (1782)

One of the connection, William Reynolds (1758-1803), who was for many years an owner of a share in the Company and who served as a kind of resident engineer, studied extensively in the various fields of science. He knew many of the eminent mechanicians and scientists of his time and experimented with electricity and chemistry as well as with foundry matters, the geology of coal, and steam engines. He also had views on education which, without mention of Bacon's name, read remarkably like those of the creator of *New Atlantis*. Writing to one of the connection, he remarked *a propos* of the joys of research into the electrical properties of the Leyden jar:

> I heartily join with thee in thy astonishment that no more young people discover a propensity to make disquisitions into these phenomena which present such inexhaustible funds of pleasure and instruction

95

& I look upon it to be a good deal owing to the faults of Education—Children are too often taught to construe a Latin book & write a good hand without ever being made acquainted with the most useful Truths of natural Philosophy which are far better suited to their Capacities & far more agreeable to their inclinations, than droning for years over a Latin Accidence which is too often the Case—the knowledge of Things is too much disregarded while that of words is too much attended to & which is the most useful as well as agreeable every one will readily determine.[23]

That intuitively Baconian philosophy would seem to have been pervasive in the Darby clan.

THE DARBY CONNECTION
AND THE DIFFUSION OF THE COKE-IRON PROCESS

In the preceding section we examined the centripetal aspects of the Darby connection, as it centered on and fed into the Coalbrookdale Company and its affiliates. Now let us look at the centrifugal aspects, or what may be termed metaphorically the seeding process, by which members of the Darby connection went out of the Coalbrookdale area to found, or to participate in the founding of, new iron companies in the British Isles. Again, the reader is reminded that this will be only a minimal sketch, other instances certainly have occurred for which no record is available to the writer; also, the Darbys are only the most conspicuous example of a widespread phenomenon.

The first of the Darby offshoots of interest involves Bersham Furnace, which was located in Broseley, just across the Severn from Coalbrookdale (it was these two towns that the iron bridge joined in 1779). The first of Abraham

[23] Quoted in Raistrick 1954, 93.

Darby's daughters, Anne, married a nail-maker named John Hawkins who acquired a share in the Coalbrookdale Company through his wife and with the profits from its sale purchased Bersham Furnace in 1727. The furnace had been built ten years before by Charles Lloyd of Dolobran, a member of another great family of iron, and in 1721 had been converted to coal. Hawkins ran into financial trouble, however, and Richard Ford and his partners at Coalbrookdale took it over, keeping Hawkins on as manager. Hawkins was technically sophisticated and developed a technique for casting pots directly out of the blast furnace without need for remelting. Bersham Furnace supplied the local country trade in pots for some years but in 1753, as the company expanded its operations northward to Horsehay and Ketley, the partners sold Bersham to Isaac Wilkinson, the Unitarian ironmaster, who promptly proceeded to use it as a gun foundry. He patented a new piston-force pump for the blast and a method for casting cannon in a dry sand-boxed mold. This was an adaptation of the previously secret Darby method of casting patented fifty years before. About 1761, Wilkinson's two sons, John and William, took over Bersham works, and they then opened a large trade in cannon, shells, and grenades with the Office of Ordnance, and later undertook there (and at other works) to become the principal machine parts maker for Boulton and Watt (as the Darbys had been for Newcomen). Bersham Iron Foundry earned the dubious reputation of being "the most considerable one of its kind in Europe. . . . Here engines of mortality of all descriptions are cast, not only for this kingdom, but for most others." Although the Quaker Darbys would not cast guns at Coalbrookdale, and refuse to employ Joseph Priestly's son because of his association with the flamboyant gunmaker and "king of the ironmasters" John Wilkinson (who had married Priestly's sister), they were not averse to direct business relationships. Finally, for a number of years, until its dissolution in 1795, one of the Reynolds served as the last manager of Bersham.

The example of Bersham Iron Foundry is instructive. It was linked with the Darby connection not only by geographical proximity but also by the presence of Abraham Darby's son-in-law as owner or as manager at least from 1727 to 1753, and by the presence of his great-grandson as manager from 1785 to 1795, with a Darby presence perhaps in the years in between. It linked the Darby connection in business and social relationships with two other great families, the Lloyds of Dolobran in Wales, and the Wilkinsons, father and sons. John Wilkinson was in turn closely tied to Boulton and Watt, and his brother William has been said to have been the expert in iron who more than any other communicated British iron and steam technology to France and Germany. At Bersham itself, during Hawkins's tenure, advances were made in the improvement of casting coal-iron that moved in the same direction as, and may have contributed directly to, the making of coal-iron suitable for forging; and Isaac Wilkinson, in adapting the Darby sand-mold technique, opened the path for the casting of cannon from coal-iron—an application not developed by the Darbys, but soon to be instituted, with the help of another connection, at the Carron Works in Scotland.[24]

The Carron Ironworks near Glasgow were established in 1759 by Dr. John Roebuck, a successful Birmingham physician and industrial chemist, and his partner Samuel Garbett, a Birmingham merchant, the country's largest importer of Swedish iron. Roebuck and Garbett, after making their fortunes in the manufacture of vitriol by an improved process using vessels of lead instead of glass, launched their iron enterprises on a large scale. They studied the methods in use at Coalbrookdale, planned the works after Coalbrookdale, and hired a Coalbrookdale man as their outside superintendent (and other men from Coalbrookdale as well). This superintendent was, as we have

[24] The Wilkinson saga is told in many places; the most thorough accounts are Dickinson 1914 and Randall n.d.

noted, one of the Darby connection, trained both at the Dale, where he had been outside superintendent, and at Bersham; his name was Robert Hawkins, and he was the son of the Bersham manager. The hard-drinking Thomas Cranage was hired to supervise construction of the first furnace. James Watt was induced to live nearby (and would later be a partner of Roebuck's before he joined with Boulton in his Birmingham days), and other distinguished engineers were brought in to help build the Carron Company into one of the largest ironworks in Europe. Carron smelted its iron with coke and introduced other major inventions such as Smeaton's cast-iron pistons to increase the blast; Smeaton's newly designed boring mill at Carron tried, and failed, to produce accurate cylinders for Boulton and Watt's steam engines. The Darbys also failed in this. It remained for Wilkinson to succeed. Carron's specialty—like Bersham's—was gun-founding; its guns, like Bersham's, were shipped to Spain, to Russia, and, after some flaws in the casting process were corrected, even to the Office of Ordnance, which finally became a principal purchaser of its light naval gun, the carronade. The great success of the Carron Company, which survives even today, was based on its ability to mass-produce cast-iron cannon of high quality made with coke-iron.[25]

The Darby connection thus accomplished the penultimate extension of the coal-iron paradigm. The final step was the smelting of iron with raw coal. This was achieved twice, more or less independently (and independently of the Darbys too), first by Robert Neilson in Scotland with the hot blast furnace patented in 1828. The hot blast made it possible to smelt iron with raw coal.

CONCLUSION

In his review of the role of Quakers in science and industry in the seventeenth and eighteenth centuries, Arthur Rai-

[25] See Campbell 1961; Birch 1955; Raistrick 1954, 149, 170; Ashton 1926, 48-52.

strick pointed out that the increasingly endogamous tendency of Quaker marriage practices itself led inevitably to dense kin networks. He referred to:

an unusually complete connection by marriage which spreads throughout the Society to such an extent that 'Quaker cousins' have almost become a byword. . . . The Quaker industries became a close network of concerns tied together by family relationship. No small business stood alone, but was helped over a difficult time by its numerous 'cousins.' The linkage in the case of the ironworks is so close that at first sight it would appear as though it were the normal policy to marry daughters to eligible small ironworks, and so bring them within the family orbit, much as estates were consolidated by marriage in other levels of society.[26]

He goes on to elaborate the industrial specializations of various regional connections—the Fell-Rawlinson group of forges in Westmorland and Lancashire, and the Lloyds of Wales and Birmingham, from about 1550 on to the end of the eighteenth century continuously carried on the process of consolidating furnaces, forges, and (eventually) collieries into combined operations; the Darby-Reynolds connection of Shropshire and south Wales, which as we have seen was responsible for the energy shift from charcoal to coke; the Cotton-Fell group in Yorkshire, which used rolling and slitting mills to produce nails, card-teeth, and cutting tools from charcoal-forged bar iron until they were put out of business by the coke-iron industry in the 1750s. Raistrick argues that these Quaker firms, working in "innocent" industries (as opposed to the ordnance suppliers) embraced the entire evolution of the English iron industry, from bloomery to blast furnace, from charcoal to coke, and from small, isolated units to large, complete-process organizations.

[26] Raistrick 1950, 44-45.

The family connection as a mode of industrial organization was not peculiar to Quakers. In the iron industry it was employed by other non-conformist families and by Catholics, Anglicans, and Methodists as well. Perhaps, because of their endogamy and their small numbers, the Quaker family connection may have been easier to mobilize in support of specific enterprises than among entrepreneurs of other persuasions. But the multigenerational family partnership, in effect managing an estate for a larger non-unilineal descent group and increasing its capital by judicious marriage alliances, was conventional enough among the gentry and aristocracy. It was simply being appropriated as the preferred mode of capital formation and industrial management, perhaps by the burgeoning industrial middle class generally, and certainly by the ironmasters.

The importance of this kinship structure to technological innovation, therefore, lies not in its uniqueness to the Darbys or to the Quakers. It was common. It provided, however, the only way in which, in the iron industry at that time, a stable cadre of mechanicians and managers could be assembled and perpetuated over several generations, passing on the paradigm from one cohort to the next. In the close and intimate setting of the family firm, father could pass on to son, or son-in-law, the principles, the techniques, and the awareness of unsolved problems, by that combination of visual and tactile communication that is crucial in technological thinking and communication. It provided, as did the Ordnance Office in Lambeth, enduring physical facilities, with enough redundancy so that a niche for experimentation was always available, safe from the demands of day-to-day operations. The significant thing about the Darbys and coke-iron is not that the first Abraham Darby "invented" a new process but that five generations of the Darby connection were able to perfect it and develop most of its applications.

III

The Ventilation of
Coal Mines

As in other matters having to do with the social context of technological change, Bacon had thought about the subject of heroes. At the conclusion of *New Atlantis*, Bacon, ever mindful of the political importance of the ceremonial display of power, and ever eager to suggest that knowledge *is* power, described the "ordinances and rites" of Bensalem's central institution, Salomon's House. Two galleries were devoted to honoring invention: one was the depository of plans and models of "the most rare and excellent inventions"; "in the other," the host declared, "we place the statues of all principal inventors." He went on to list the principal figures and to describe the public policy of the kingdom:

> There we have the statua of your Columbus, that discovered the West Indies: also the inventor of ships: your monk that was the inventor of ordnance and of gunpowder: the inventor of music: the inventor of letters: the inventor of printing: the inventor of observations of astronomy: the inventor of works in metal: the inventor of glass: the inventor of silk of the worm: the inventor of wine: the inventor of corn and bread: the inventor of sugars: and all these by more certain tradition than you have. Then have we divers inventors of our own, of excellent works; which

103

since you have not seen, it were too long to make a description of them; and besides, in the right understanding of those descriptions you might easily err. For upon every invention of value, we erect a statua to the inventor, and give him a liberal and honourable reward. These statua's are some of brass; some of marble and touchstone; some of cedar and other special woods gilt and adorned: some of iron; some of silver; some of gold.[1]

In this list, after two maritime heroes, the third place is occupied by "the inventor of ordnance and of gunpowder"; in other writings, in naming the greatest inventions, Bacon also mentions that great aid to navigation, the magnetic compass. His emphasis on naval and ordnance matters, however, natural enough in an English philosopher of his time, does not overwhelm his interest in the broad range of industrial processes.

There is an interesting omission in Bacon's account of how Bensalem rewarded its inventors: he makes no explicit mention of the granting of a patent of monopoly to inventors. This may be due to the fact that Parliament in 1624 had disallowed most patents of monopoly (except inventions); but Bacon was probably not in favor of that action, and he did envision some such patents being granted to the family corporations of Atlantis, as we saw earlier. The failure to grant patents to inventors may actually reflect Bacon's own vision of the progress of science and the arts as a collective enterprise. In one place he compares it to a race run by teams of runners, each passing on the torch to his successor;[2] in another he declares that the "arts and sciences should be like mines, where the noise of new works and further advances is heard every day."[3] Thus, although the scientist and inventor deserved

[1] Bacon, *Works*, 1: New Atlantis, 411-412.
[2] As quoted in Rossi 1968, 10, from the *Wisdom of the Ancients*.
[3] Bacon, *Works*, 1: Novum Organum (x c), 127.

the reward of fame (Bacon compared himself, indeed, to the same Columbus who held first rank among the heroes of Salomon's House[4]), he was still a public servant whose innovations belonged to the commonwealth.

Bacon thought much upon social control, about the means by which men are motivated to act for the good of the commonwealth. Honor, fame, reputation: he saw these as quanta of social approval that men sought universally, but not necessarily wisely, to maximize. In the essay, "Of Honour and Reputation," he remarks on the criterion of novelty as important in maximizing honor:

> If a man perform that which hath not been attempted before; or attempted and given over; or hath been achieved, but not with so good circumstance; he shall purchase more honour, than by effecting a matter of greater difficulty or virtue, wherein he is but a follower.

And he adds that there is another criterion—the degree of risk:

> There is an honour, likewise, which may be ranked amongst the greatest, which happeneth rarely; that is, of such as sacrifice themselves to death or danger for the good of their country; as was M. Regulus, and the two Decii.[5]

In Bacon's time, common men were not ordinarily eligible for honor; it was primarily reserved for nobility and for the court. For him to suggest, in *New Atlantis*, that mere artisans would in a well-ordered realm be eligible for the highest honors seems to press the bounds of possibility in the England of Charles I; it could happen only in utopia.

Yet within three decades, a Royal Society existed, under charter of Charles II, avowedly inspired by Bacon's vision

[4] Levine 1970, 78.
[5] Bacon, *Works*, 2: Essays, 263-265.

of the scientific commonwealth. Within the century, knighthood was conferred upon English scientists; within a century and a half, it was being offered to outstanding technological innovators too (Richard Arkwright accepted, and James Watt refused, the honor). But the real apotheosis of the industrial hero came more than two hundred years after the publication of Bacon's *New Atlantis*, when in the mid-nineteenth century his fellow countryman, Samuel Smiles, deliberately touted the engineer as the risk-taking benefactor of mankind.

Samuel Smiles was a physician with experience as a railroad official, who wrote his popular lives of the engineers to celebrate the new kind of hero who was remaking the world. Smiles viewed the engineer as a man who, by developing new sources of energy, by inventing new machines, by opening new lands to cultivation, and by devising new means and channels of transportation, opened up benighted regions to the light of modern civilization and brought order and health to ignorant, lawless, hungry, and diseased millions. Such benefactors of mankind were a special breed, emerging from the ranks of the rural working class, largely self-taught, and nobly courageous. The first exemplar of this type to receive Smiles's treatment, in 1857, was George Stephenson, coal miner, mining engineer, inventor of a successful miners' safety lamp, and famous as the developer of one of the first successful steam locomotives. Smiles's rhetoric in describing Stephenson's first trial of his safety lamp, as he advanced alone into a gassy mine with his lighted lamp in hand, defined the character of the industrial hero in vivid terms and explicitly contrasted him with the military hero:

> Stephenson declared his confidence in the safety of his lamp, and, having lit the wick, he boldly proceeded with it toward the explosive air. The others, more timid and doubtful, hung back when they came

106

within hearing of the blower; and, apprehensive of the danger, they retired into a safe place, out of sight of the lamp, which gradually disappeared with its bearer in the recesses of the mine. It was a critical moment, and the danger was such as would have tried the stoutest heart. Stephenson, advancing alone, with his yet untried lamp, in the depths of those underground workings, calmly venturing his life in the determination to discover a mode by which the lives of many might be saved, and death disarmed in these fatal caverns, presented an example of intrepid nerve and manly courage more noble than that which, in the excitement of battle and the collective impetuosity of a charge, carries a man up to the cannon's mouth.[6]

Smiles's book was published in an American edition in the same year.

Smiles went on to write competent and readable biographies of other engineers, including the seventeenth-century figures Cornelius Vermuyden and Hugh Myddelton, who promoted the drainage of the Fens and London's water supply; eighteenth-century mechanicians James Bradley and John Smeaton, canal builders and improvers of the steam engine; Abraham Darby, who first successfully smelted iron with coke; and nineteenth-century bridge and canal-building engineers John Rennie and Thomas Telford. He wrote books on the improvement of character with titles like *Self-Help*, *Duty*, and *Character* (and cited Bacon as a courageous innovator). Throughout, he emphasized the goal of development—the development of wild, uncultivated, uncivilized lands, and the correlative development of men of character. He articulated a mystique of industrial progress, seeing it as the means by which poverty, ignorance, violence, and vice could be eradicated, and prosperity, education, cooperation, and virtue put in their place. His heroes, largely drawn from

[6] Smiles 1859, 182; see also Smiles 1966.

poor rural circumstances, developed their own characters and achieved wealth; with dauntless courage, limitless patience, and indomitable perserverance they worked to make such a good life possible for others. Smiles saw the process of economic development as a complex unfolding that required the services not only of inventors but also engineers and investors, not only of mechanicians but also of scientists. It was no mere Tom Swift the Boy Inventor fable that he was creating, but a valid myth for the times that celebrated the heroic virtues appropriate to the goal of industrialization. As confidence in industrial society itself has declined, of course, so has Smiles's standing as a literary figure; but that is another story.[7]

Nowhere was the myth of the industrial hero apparently more needed than in the coal fields of Great Britain, France, Belgium, and the United States, where the new source of energy to fuel the Industrial Revolution was being extracted from the earth at great cost in death, injury, and destruction. It is no accident that Samuel Smiles, the official biographer and chief hero-maker of the Industrial Revolution in Britain, selected a coal miner (and also engineer and railroad man)—George Stephenson—as the first subject in his series of Lives of the Engineers, for the coal mines were the front line of battle in the campaign to industrialize the world.

COAL MINE EXPLOSIONS
AND THE PROBLEM OF VENTILATION

Throughout the eighteenth century, as England's need for coal increased, and particularly after the coke-iron process came to be used more widely, English coal mines expanded in number and size. In America, a similar development of coal mining began in the 1840s, after it was discovered that iron could be smelted with pure Penn-

7 Smiles 1857, 1877, 1881.

sylvania anthracite. In both countries, however, mining remained an extremely dangerous trade in which the underground miner risked life and limb and his employer risked bankruptcy. One of the principal causes of death and destruction was the explosion of fire-damp, or methane gas, which constantly exuded from the coal seams. Those who worked in these explosive mines, who invested their fortunes in them, or who invented new systems of ventilation and lighting to prevent catastrophe, could properly be called heroes of the Industrial Revolution. Nowhere, indeed, was the heroic inventor more needed than in the development of improved systems of ventilation and lighting for gassy mines.

Fire-damp, or methane (CH_4) was more or less constantly exuded from coal seams in varying quantities; at times, however, large pockets of gas were broken into, resulting in "blowers" that released large volumes very rapidly. Most explosions occurred in individual stalls, involved only local pockets of gas, and killed or burned only the miner and laborer working in that stall and perhaps adjacent mule-boys, passing miners, and others nearby. In dry mines with floors laden with inches of finely powdered coal dust, however, the initial gas explosion could stir up a cloud of coal powder that also ignited. As the expanding fire-ball of incandescent dust and gas stirred up still more dust, the entire mine turned into "an enormous piece of artillery" (to use Sir Humphry Davy's phrase), hurling coal cars, mules, and men down the gangways and even up and out of the shaft. Hundreds of men died in each such major explosion. Even if this ultimate catastrophe did not occur, sufficiently prolonged burning of gas and dust could start an underground fire that burned timbers, ignited the coal, destroyed the shaft under the hoisting and ventilation system, and asphyxiated all who remained below ground. Some such fires burned for years, resisting all efforts to put them out by flooding the mine or sealing the airways.

The problem of preventing coal mine explosions had drawn the attention of technologists ever since the mid-seventeenth century, when English coal mines began to expand in size. No longer shallow, bowl-shaped pits, sixty feet or so in depth, they became a complex network of galleries driven eventually for miles through veins reached by deep shafts. Under these conditions, underground explosions began to occur. It was well known that, if adequate ventilation could be established, noxious vapors (including explosive fire-damp) could be exhausted from the mine. Agricola's classic treatise on mining, published in Latin and German in 1556 (and well known to Bacon) described various devices, including large horse- and water-powered centrifugal blowers, for introducing fresh and sucking out poisonous air from the precious metal mines in central Europe.[8] But these mines did not ordinarily generate the explosive fire-damp. In English coal mines, on the other hand, explosions had become frequent enough by the middle of the seventeenth century for the problem of mine ventilation to be repeatedly drawn to the attention of the Royal Society in its first decade. In the context of a series of "subterraneous experiments" on varying barometric pressures in coal pits, Henry Power, a Cambridge-trained physician and a frequent participant in the Society's meetings, presented in 1662 a paper on three damps in coal pits and the methods of ventilation then practiced, which he illustrated by a schematic diagram of typical underground workings (see fig. 16). He noted that fire damp explosions were known to have killed and mangled men in Newcastle pits. The principal method of ventilation was what later came to be called "natural ventilation":

Every coal-pit hath its vent-pit, digged down at a competent distance from it, as 50 or 80 paces one from another. They dig a vault under ground, from one pit to another, which they call the vent-head,

[8] Agricola 1950, 200-212.

110

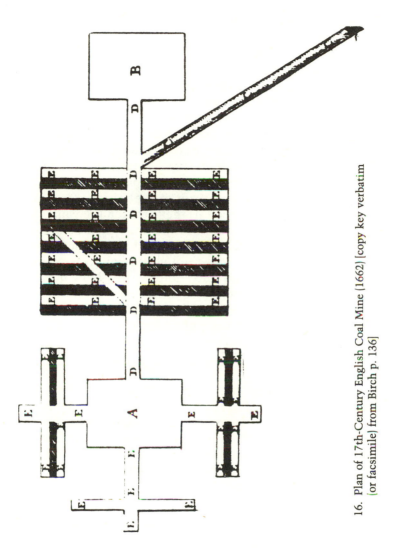

16. Plan of 17th-Century English Coal Mine (1662) [copy key verbatim (or facsimile) from Birch p. 136]

that the air may have a free passage from one pit to the other. Now the air always has a motion, and runs in a stream from one pit to the other; for if the air should have no motion (or vent as they call it) but restagnate, then they could not work in the pits. It is not requisite, that the vent-pit should be as deep as the coal pit. Now the vent, or current of subterraneous air, is sometimes one way, and sometimes another; sometimes from the vent-pit to the coal-pit, and sometimes contrariwise, as the winds above ground do alter; and also weaker and stronger at some times than at other; and sometimes the vent plays so weakly, that they cannot work for want of ventilation. Then, to gather vent (as they call it) they straiten the vault, and wall part of it up; so that the air, which before ran in a large stream, being now crowded into a lesser channel, and forced to pass through a narrower room, gathers in strength, and runs more swiftly. Now it is observed, that the subterraneous air is always warm, and in the coldest weather, the warmest; so that it never freezes in the pit out of which the vent plays.

He also noted that, in order to increase the current of air, artificial means were sometimes employed:

Now besides the playing of the vent, they sometimes are necessitated to keep constant fires underground, to purify and ventilate the air; sometimes the running of the scoops, when they begin to work, will set it in motion; sometimes, if the damp draw towards the eye of the pit, they set it into motion, by throwing down of coalsacks: else the heavy vapour will restagnate there, and is not able to rise.

Thus by 1662 the technique of "furnace ventilation" was already in use in English coal mines.[9]

[9] Birch 1756-57, 1: 133-136. Merton 1970, pp. 148-150, discusses the early concern of the Royal Society with mine ventilation at some length.

One of the selling points of Savery's steam engine, as he forcefully stated in *The Miner's Friend*, was that it improved the ventilation. Remembering that Savery's engine, dependent upon atmospheric pressure to raise water to the boiler, had to be placed no more than twenty feet or so from the sump at the bottom of the pit, we may read his words as another proposal for placing a furnace in the mine shaft to improve ventilation:

Miner. But we have often combustible vapours in our mines, which taking fire from the candles used there, do, by a sudden explosion, destroy both the mine and the miner; and therefore I am afraid that the fire used in your engine will be very dangerous, and apt to kindle those combustibles more than our candles.

Author. To answer this objection, I will desire leave to give you my notion of those combustibles, which, in short, is this: that when your miners come to a close place, where there is no circulation of air to carry off the effluvia, or atoms constantly rising like fine dust in a powder-mill, they by knocking and working do increase to be very numerous, like to those loose particles in a powder-mill. But it is the work of some time for those vapours to come to perfection; for I have heard several experienced miners say, that it is common to avoid the danger of those vapours, by retiring as soon as they see the flame of their candles burn longer than ordinary, which may be discerned sometimes long before the air is thick enough of this combustible matter to take fire at once, and, like gunpowder, to destroy all. I did hear one say, that from an inch and a half, once the flame of his candle did gradually increase to two feet long, and yet he escaped. Which makes it very plain, that stagnation of air is the sole cause of this inconvenience in mines, which may be totally prevented by a pipe going from

113

the ash-pit of our furnace to any part of the mine liable to stagnation. For the air will press with great violence through the pipe into the fire, before the combustible matter can be ready to do any hurt, and passing through the fire, make way for fresh air to descend in the room of it. So that our fire, instead of blowing up your works, is the best means that can be used to prevent so fatal an accident; and will likewise carry off all unwholesome vapours, damps, or steams, which may proceed from corruption of air, by stagnations or vapours arising from any poisonous earth or mineral.[10]

Probably by the mid-eighteenth century, it was common practice in English mines to build a furnace (without Savery's engine attached to it) at the bottom of the mine shaft, to carry up through a chimney—in the best designed mines, a separate shaft, and otherwise a partitioned-off portion of the shaft—the dangerous air from the working gangways.

But these measures did not suffice. About 1814, the renowned English chemist and Fellow of the Royal Society, Sir Humphry Davy, responded to a call for help from a committee of coal mine proprietors in the Newcastle area after a gas explosion in one of their collieries killed 101 men. Davy, aided by his ingenious laboratory assistant Michael Faraday, first analyzed the chemical composition of "fire-damp," as the gas was then called, and found, as others had before him, that its formula was CH_4—the same colorless, odorless, lighter-than-air gas now called methane, a principal constituent of the natural gas so widely used for domestic heating and cooking today. Advised that most fire-damp explosions were set off by contact between incautious miners' lanterns and pockets of gas in the underground workings, Davy and Faraday proceeded to invent a safety lamp. After trying, and failing,

[10] Savery 1702, 49-50.

114

to devise a satisfactory electric light, they produced the famous Davy lamp, which separated the flame from the surrounding mixture of fire-damp and air by a heat-diffusing metal screen. As we have seen, George Stephenson—and, in fact, a number of others as well—devised similar safety lamps. Davy's new lamp flared up but did not explode in dangerous atmospheres; Stephenson's was extinguished. Davy's new lamp in particular proved to be much more popular than older safety lights and, despite the feebleness of its beams, was used in the most dangerous mines; Davy was widely heralded as the savior of the coal trade. He concluded his book of 1818 describing his researches with an optimistic testimonial to his faith in the efficacy of rationally enlightened self-interest in promoting both private profit and public good:

When the duties of men coincide with their interests, they are usually performed with alacrity; the progress of civilization ensures the existence of all real improvements; and however high the gratification of possessing the good opinion of society, there is a still more exalted pleasure in the consciousness of having laboured to be useful.[11]

But the effect of the introduction of the safety lamp turned out to be, in actuality, a *rise*, not a decline, in the number of fatalities from gas explosions in English coal mines. Operators were encouraged to send miners into workings previously considered too dangerous to enter; and miners, impatient with the lamp's poor light, too often used a naked candle instead, testing for fire-damp by observing the length of the flame.

Thirty years later, after a long series of disastrous coal mine explosions, Parliament called on Michael Faraday, now famous for his studies in electrical phenomena, and geologist Sir Charles Lyell to investigate the problem once

[11] Davy 1818.

115

more. Faraday's report was published in England in 1844 and an abstract was printed in the *Franklin Journal* in Philadelphia in 1845. Faraday found again that failure to use the safety lamp, and failure to provide proper ventilation, caused gas explosions. But he returned to the problem of the human agent. No longer was he satisfied (if he had ever been) with Davy's hopeful assumption that the calculus of private gain was the same as that for the public good. He went further and questioned whether even the private motive of self-preservation was sufficient to prevent accidents. He had a grimly humorous anecdote to tell about himself to illustrate the point. In order to understand the process of mining better, he said, he spent time following the miners underground, watching them work, and noting with dismay their careless habits of neglecting the Davy lamp and smoking in potentially gas-laden areas. One day he was watching a small group of miners drilling in a face of coal in preparation for blasting with gunpowder. Faraday sat on a convenient keg watching the preparations with a lighted candle in his hand. When drilling was completed, Faraday inquired where they had put the gunpowder and was told that he was sitting on it! Faraday's opinion as to the possibility of reducing the frequency of accidents was somber:

> But the great source of danger was the mental condition of the miners. With regard to the present race this is so hopeless that nothing could be done for them . . .[12]

Faraday was pointing to a peculiar feature of the risk-taking behavior of English miners and mine operators: not merely did they risk, and lose, other people's lives, income, and property in the pursuit of private gain; they risked and sometimes lost their own, too. The invisible

[12] *Journal of the Franklin Institute*, 1845, part 2: 287-288.

hand of enlightened self-interest seems to have trembled a little when it reached into the coal fields.

The working miner's responsibility in preventing fire-damp explosions was primarily the negative one of not carrying a naked lighted candle or unshielded oil lantern into a pocket of fire-damp. When he, or the fire-boss, suspected the presence of fire-damp, he was supposed to use the Davy safety lamp or one of the many modifications of it on the market by the 1850s; but awareness of the odorless, colorless gas might come too late. The protection provided by the safety lamp (and later by electric lights and battery operated miner's lamps) was rendered even more uncertain by the fact that there were other sources of ignition—matches, fuses, gunpowder explosions, and percussion sparks. The basic responsibility for preventing fire-damp explosions fell, everyone agreed, on the colliery operator. The mine had to be so designed that a constant flow of air across the face of all coal seams constantly carried off the fire-damp and replaced it with fresh air. To accomplish this successfully, in any but the smallest works, two separate shaft openings were required—the main working shaft and an "air hole"—one for the downdraft and one for the updraft. Under some optimal conditions, adequate ventilation could be achieved by natural means, the direction of the flow depending upon whether the air in the mine were warmer or cooler than outside. But for practical purposes, in a deep mine with a series of gangways one on top of the other, extending in total length ten miles or more underground, artificial means to move the air were required. As we have seen, in deep English mines, from early in the eighteenth century, the ventilation was increased by building a furnace at the bottom of the upcast; as the heated exhaust from the furnace ascended, fresh air entered the mine through the separate downcast. By the time the "fresh" air reached the furnace, it was laden not only with carbon dioxide but also with all the fire-damp it had collected along the way. This mixture could ignite in the furnace and flash back along the gangways. So in the better-designed mines a "dumb drift" was tunneled through the rock to bring most of the noxious air into the upcast above the furnace itself. Chimneys were often built on the surface above the upcast in order to aid the draft. After about 1860, steam-powered centrifugal blowers gradually replaced the furnaces as the prime air mover; they were usually constructed over the upcast and functioned as enormous exhaust fans. (They followed Papin's basic design of the 1690s which in turn was merely an improvement of the German centrifugal fans illustrated by Agricola in the 1550s.)

Providing a source of fresh air was only the beginning of the ventilation problem. Throughout the entire mine, an elaborate system of coal pillars, wooden partitions, and doors had to be constructed to conduct the air where it should go. Any design error, as well as any accidental opening of a door or collapse of a partition or fall of rock could invalidate the ventilation system in any part of the mine. Finally, the architectural problem in designing a mine was not simply a matter of initial planning; as the mine evolved, with gangways added here and whole sections abandoned there, its ventilation needs changed and its problems shifted from place to place.

INNOVATION IN MINING TECHNOLOGY:
LIGHTER-THAN-AIR GASES AS VENTILATORS

The Newcastle mine proprietors took the lead in encouraging the development of the safety lamp. Although aiding ventilation by fires or furnaces at the bottom of shafts or at the mouth of adits was occasionally practiced in Belgium and in England in the mid-seventeenth century, well before Savery's suggestion, the practice was later on evaluated by systematic investigation in the Newcastle area and declared to be the best practice for ventilation, superior to the steam-jet (another lighter-than-air procedure). The Newcastle coal field was worked by large landowners whose estates had other sources of revenue than coal; careful and deliberate investment in the best practices, and in research and experimentation to improve practice, made it possible for them to exploit the deep, and relatively flat, seams of coal that underlay the region. It was in the coal districts of the north of England, for instance, that one of the most systematic methods of laying out the mine, the "board-and-wall" system, was developed.[13]

A different situation prevailed in parts of Staffordshire. Here coal outcropped on the surface and could be and was mined in small pits by large numbers of ill-financed operators, each anxious to reap his reward in cash as fast as possible. As the Newcastle-based author of an 1851 treatise on mine ventilation explained, this situation was inimical to best practice:

That haste, which is common among inexperienced adventurers in the coal trade, to raise large quantities of coal, in an incredibly short time after the shafts are sunk, cannot be too strongly condemned: such persons cannot see why they should not at once commence raising large quantities, and think that others are supine who do not, as soon as the shafts are sunk, begin a wholesale raising of coal. The experienced coal owner or manager is however aware that there is some preparation absolutely necessary, before this can be done with propriety. This hasty wholesale raising of coal before the necessary preparations are made, is frequently attended with disastrous results. . . .

In some mining districts the Proprietor lets the working of his mines to a Contractor (in Staffordshire called a Butty), who provides all labor to raise the coal

[13] Nef 1966, 1: 357, 363-364; Daddow and Bannan 1866, 426-438; Atkinson 1892.

119

and to maintain the mine in a workable state. The Contractor again sub-lets the different branches of labor to the workmen. This way of letting mines operates to a great extent prejudicially to the introduction of good ventilation and discipline; almost every man's object being, under such circumstances, to make the most of his bargain, he attends to such of his duties as bring him a return for his labour, until urgent necessity requires the contrary, and thus, provisions which are necessary for the safety and welfare of the mine are neglected. I admit there may be Contractors who feel the responsibility that rests upon them, and provide not only good ventilation, but conduct the mine under as good regulations as circumstances allow. But the practice being bad, the sooner it is abolished the better, and in its place an approved system of management introduced, whereby the mines may be placed on a footing more secure from accidents, and the produce be raised with greater economy.[14]

Nevertheless, there were in Staffordshire and adjoining Worcestershire some large proprietors who mined their estates extensively and innovatively, and the most famous of these were the Dudleys of Dudley Castle. We have already noted Dudley Castle as the site of the first Newcomen engine, set up in 1712 to pump water out of the coal mines; and we have also made note of Dud Dudley's partially successful early efforts from about 1619 to 1663 to develop a process for smelting iron with coal. From the late sixteenth century well on into the present century, the Earls of Dudley and their kin maintained very extensive operations in coal mines, furnaces, and foundries on their lands in Worcestershire and Staffordshire. These are described for the early period in Dud Dudley's book of 1665, *Metallum Martis*. Staffordshire had strip

[14] Hedley 1851, 20-21.

120

mines and sloping pits as deep as 120 feet. By the middle of the eighteenth century, the Dudleys had gone far beyond simple pit and strip mining and were driving extensive galleries underground. They continued to be innovative. From 1769 to 1792, the Dudleys invested in feeder canals to join their collieries to the Birmingham and Stourbridge Canals, including a 3,142-yard-long tunnel—the Dudley Canal Tunnel—through their collieries half a mile from Dudley Castle. This tunnel, supplied with air vents to the surface, undoubtedly contributed to the ventilation of the mines through which it passed. The tunnel was in use until the 1950s.[15]

The deeper mines of Staffordshire, because of the steeply inclined angle (on the order of 60° to 80° from the horizontal) of the outcropping coal measures, were reached by slopes (to use the American term). A slope was sunk by digging a sloping pit down through the coal to a sufficient depth, and then driving gangways horizontally through the coal, with rooms or stalls opening off the gangway upward to the coal. (In Newcastle, by contrast, the coal was reached by vertical shafts driven through rock.) At some point in the history of the Staffordshire mines, it was discovered that fire-damp itself, being considerably lighter than air, could be used instead of air heated by a furnace to form an ascending air column.[16] In one application of this principle, described in a work published in 1859, the return airway was driven through the thirty-foot coal seam directly above and parallel to the gangway. A "spout" was dug perpendicular up from the gangway to the airway every fifteen yards, through which the methane escaped. Where too much gas accumulated to permit men to work in the spout, a four-inch hole could be drilled, and it served the purpose beautifully.[17]

English innovations in mining technology, and their

[15] Dudley 1665, 26-2?; Sarjeant 1964.
[16] Daddow and Bannan 1866, 441.
[17] Moore 1859.

principles of best practice, were exported to America in the 1830s and 1840s, when the great development of the Pennsylvania mines was taking place. Welsh and English miners, operators, and engineers flocked to the new fields, and particularly to the anthracite district, about ninety miles north of Philadelphia. Anthracite, or hard coal, was regarded as scarcely worth mining in south Wales, where the British deposits lay, and it was American innovations that made the development of anthracite measures profitable. In the mid-1820s, newly designed hearth-grates and instruction in their use made anthracite practical as a domestic fuel; its cleanness (being nearly pure carbon, it produced virtually no odor or soot) made it more attractive than soft coal, and diminishing supplies of wood had driven up the price of firewood in urban areas. In the 1840s, techniques for using anthracite directly, without coking, as a superior substitute for charcoal in blast furnaces contributed to the rapid increase in Pennsylvania's iron and steel industry. Thereafter, in the anthracite area, one innovation followed another in the rush to increase production. In 1844, the first of the new steam-powered mechanical coal breakers invented by Joseph Battin was put in operation; thereafter the huge wooden breaker houses where the lumps of coal from the mines were broken and sorted into sizes began to dominate the landscape at the

17. Methods of Sinking Coal Mines in America (1866) a. slope b. shaft c. drifts and tunnels d. gangways e. breakers

mines. New elevators, new exhaust fans, new Cornish bull steam engines, new diamond drills, new water pumps came on in an endless parade of mechanical ingenuity.

Most of this ingenuity was expended upon above-ground mechanical equipment. Although a great deal of concern was also expressed about ventilation, the institutional setting for best practice in ventilation was absent in most mining operations in Schuylkill County, for some years the principal producer of coal in the anthracite district. That best, and even innovative, practice was not impossible, however, can be shown in the career of a Welsh mine operator, William H. Johns, and his family.

Johns was born in Pembrokeshire, in south Wales, came to America in 1832 at the age of 17, and prospered in the mines of Schuylkill County. At his death in 1865, he left a fortune of over a million dollars. He, and later his son,

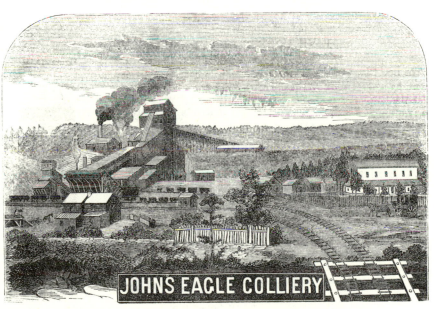

18. Johns' Eagle Colliery (1866)

123

developed a single mine in the little town of St. Clair from a small drift to a truly gigantic colliery with over fifteen miles of gangways open in 1869, a few years before it was sold. It was then described by the state's mine inspector as "the most extensive colliery in the Union."[18] As descriptions in various sources reveal, part of the Johns's success lay in their constant adherence to principles of best practice; and not least in their regard was the ventilation system. Apparently they followed a modification of the ventilation system used in Johns's native Pembrokeshire, where the seams were tilted in a manner similar to that in Schuylkill County. In the Pembrokeshire system (which may in fact have been the inspiration for, or else borrowed from, the Staffordshire mode), much of the work of ventilation was done by venting the methane upward through sloping headings cut through the coal vein that connected a lower main gangway and air inlet with an overlying parallel return air course. The stalls were opened off these headings and they in turn were connected by a series of air headings so that the lighter-than-air fire-damp was exhausted directly upward from the working.[19] Johns's initial modification of this technique was to drive airways to the surface (about 80 yards) from the gangway before mining began, and then to work the stalls (or breasts, as they were called in Pennsylvania) all the way to the surface, thus progressively leaving behind, as the work advanced, a series of air courses through which much of the fire-damp could be exhausted from nearby working areas.[20] As the mine evolved, Johns sank a slope on the forty-foot thick Mammoth Vein, and this he may have designed on the south Wales plan, for as late as 1870 it was said to be ventilated by natural means

---

[18] *Report of the Inspector of Mines*, County of Schuylkill, 1869 (Harrisburg: B. Singerley, 1870), p. 160.

[19] *Miners' Journal* (Pottsville), 7 July 1860.

[20] Schuylkill County Historical Society (SCHS), Maddison Reports, Eagle Colliery, 7 Sept. 1850 and 9 Nov. 1853.

without a furnace or even a fan. Johns's method (and that in use in Pembrokeshire and Staffordshire) was put forward under the rubric "Gas as a Ventilator" in an 1866 book on mining technology written by a friend and neighbor of the Johns family. The language is very similar to that used in describing the Staffordshire method in an English work published seven years earlier.[21]

The experience of the Johns family in the highly competitive and economically unstable Pennsylvania coal trade is evidence that innovation and "best practice" went hand in hand with—and, indeed, may have been an essential condition of—financial success. A kind of control for the Johns case is provided by a couple of the Johns's neighbors and competitors, who did not employ best practice in ventilation, and who—like many among his peers—suffered bankruptcy as a result.

REGRESSION IN VENTILATION PRACTICE:
TWO MINE OPERATORS OF ST. CLAIR

It is readily apparent that the one indispensable thing in the ventilation system of a coal mine is the separation of upcast and downcast airways. Failure to provide and to maintain such a separation immediately establishes an emergency situation in which the likelihood of a gas explosion rapidly increases to near-certainty. With this fact in mind, let us examine the behavior of two mine operators—neighbors of the Johns family—who produced anthracite in St. Clair. After examining their behavior, I shall interpret it in terms of economic motivation and also in the light of the "industrialist as hero" theme.

The first large colliery to be opened at St. Clair was Parvin's Slope. In the spring of 1852 a local operator named Francis Parvin, leasing from land owner Henry Carey (the well-known Philadelphia economic philosopher), sank a

[21] Daddow and Bannan 1866, 441-442; Moore 1859, 50-52.

slope 333 feet long in the ten-foot thick Primrose Vein. It was expected to provide at least 35,000 tons per year. Two years later, it was in fact producing 3,000 tons per month, from gangways on three levels, and in 1856 it shipped over 30,000 tons. By the fall of 1859, seven years after commencing operation, the slope had been extended to a total of 600 feet. From the beginning, troubles had been accumulating. Parvin miscalculated the angle at which the vein dipped into the earth and had to tunnel through solid rock to straighten it out at the top and bottom; one of the miners cut into a spring that constantly, and for years, threatened to drown the works; early in 1854 the workings ran into a "pinch" (a constriction of the seam) that reduced production to 500 tons for the whole year; and they had trouble with ventilation. Although an airway was sunk close and parallel to the slope (which itself was so narrow that ordinary mine cars could not be admitted), the engineer reported to Carey's agent at the very beginning his concern over the inadequacy of the ventilation

19. Slope Exposed (Artist's conception) (1857)

plans "for the present and future working of these mines."
No furnace was installed to increase the movement of air,
and finally in the fall of 1859, after sinking the slope the
last hundred feet, Parvin failed to complete driving the
airway down the last fifteen yards to the level where workmen were already extending gangways. There was thus
no circulation of air at all, and within a few days a large
explosion occurred. How many men were killed or injured
was not reported, but the fireball was sufficiently prolonged to set the mine on fire, burning the supporting
props and top timbers in the gangways, and the timbering
in the slope itself, which resulted in the entire wooden
mass falling down the hole, crushing the pumps at the
bottom. The fire burned for days until the mine was deliberately flooded to put it out. Pumping the water out
proved to be difficult because of the perennial spring. Although Parvin eventually got the water out and some of
the gangways open again, he continued to have trouble
with water, and was unable to ship any coal at all in 1860.
In 1861 he was forced into a sheriff's sale, unable to pay
off $10,000 in debts incurred in reopening. The mine itself
was abandoned forever and the machinery dispersed after
nine years of operation.[22]

The next deep mine to be opened on the St. Clair Tract
was the slope sunk in 1852 by Enoch McGinness in the
Mammoth and Seven Foot Veins. It extended from the
northeast corner of the tract southward to the level where
the Mammoth bottomed out, about 450 feet below the
surface. It produced 19,080 tons of coal in 1853, the first
year of operation, and over 41,000 the second year, but
the tonnage declined gradually after that and mining seems
to have been given up in the slope workings after 1860,

[22] The sinking of Parvin's Slope is described in the report of the engineer John Maddison (Schuylkill County Historical Society.) Its problems thereafter are recorded in the letters of Enoch McGinness to Henry
C. Carey (Carey-Gardiner Coll., HSP), and its final demise is noted in
the engineering notebook of George K. Smith (SCHS).

20. Working in Steep Breast (1866)

the slope itself and its airways being used thereafter as the intake for the ventilation of the St. Clair Shaft.

The problems with the slope were largely water problems. A spring was cut in the lower works in April 1855 that let in more water than the existing pump could throw out. McGinness installed a new, heavier pump, but it broke the main shaft of the pumping engine, and the slope was not clear of water until June. In August the barrel of the new pump burst, and another week's income was lost. In September, McGinness, who was having difficulties also with the shaft colliery, had to sell out in order to pay his debts. The saga of the St. Clair Slope goes on and on in its watery way for the next ten years. In 1858 it was flooded out again and the water threatened to break through into the shaft and flood that, too. The slope was drained, and the lower level gangways of the shaft and slope were connected to make an airway for the shaft; but now all the water from the slope that escaped the pump drained into the shaft, which promptly flooded the first time the slope pump gave out. Under these conditions, little coal could be mined at either the shaft or the slope, and the slope, virtually abandoned after 1861, deteriorated badly, timber props rotting out, airways closed by rockfalls, and the chimney-stack over the ventilation furnace blocked

128

with rubbish. By 1865, two more bankruptcies later, an engineer reported to Carey that no one could enter the slope because of black damp (carbon-dioxide) and methane. The engineer complained: "They are not trying to get air in. If it must stand [i.e., be abandoned], for fear of destroying everything else by a gas or sulphur explosion, a watchman should be kept there to prevent any person going near the top, to even strike a match."[23] A fan was installed on the engineer's instructions, and the slope began to ship coal again in August and continued for the rest of the year. But as soon as the flooded shaft re-opened in 1866, the slope was abandoned once more; its payroll for 1868 listed only two watchmen as employees.

The most spectacular, if not the most successful, of the St. Clair mines was the St. Clair Shaft. Perpendicular shafts of great depth, approaching half a mile, were common in England, but in America, where coal mining began later, deep mining had only begun with the first slopes in the 1830s, and the first shafts were not attempted until the 1840s. One of Carey's associates had sunk a shaft in the St. Clair Tract to the Primrose Vein, but abandoned it. Enoch McGinness, an aggressive entrepreneur with a total capital of four dollars and fifty cents, managed to convince Carey and others that he could drive the old shaft straight down to the basin of the Mammoth Vein (in contradiction to the theories of geologist Henry D. Rogers). With a line of credit from Carey, McGinness first sank a bore hole in 1852 that reached the Mammoth at a depth of 450 feet from grass; he then enlarged the shaft to its final dimensions of 10 feet by 18 feet in 1853 and put in the necessary pumps, hoisting machinery (of a newly patented design) and breaker, supplying his engines and some of his need for cash with coal from the nearby slope. The total cost

[23] The engineer's comments are quoted in H. C. Carey 1865, 25. A copy is located at the Library Company of Philadelphia.

129

of opening the shaft amounted to about $100,000. By the fall of 1854 he was able to make his first shipment of coal.

A year later, Enoch McGinness was in debt to the amount of $147,000, and was able to pay off this vast sum only by selling the colliery to another firm (with, one may suspect, the covert assistance of Henry Carey and others). Embittered, McGinness for a time even suffered the humiliation of being supported by his wife, who seems to have kept a boardinghouse as a hedge against hard times.

The crisis in McGinness's affairs was precipitated by two gas explosions in the shaft, compounded at the end by flooding. He did not describe the explosions in his own correspondence with his patron Carey, but Carey's business agent, Andrew Russell—a man whose conservative temperament McGinness despised—wrote Carey the details. The first incident occurred on June 11, 1855; Russell's account was terse:

> There was an explosion at the Shaft this day. . . . One man dangerously burnt (will probably die) one other severely & three slightly—no injury done the works— air good in the mines it was the result of carelessness.[24]

The second explosion, four months later, killed three miners and burned two others. The cause, again, was attributed to carelessness on the part of the miners:

> It appears that the mining boss had examined one of the breasts and informed the miner, who was engaged in working it, of the fact, cautioning him at the same time, not to go into it.—But disregarding the warning, the most reckless and unpardonable carelessness, he ascended the breast with a naked lamp flaming on his cap and an *unlit Safety Lamp in his hand.*

[24] Andrew Russell to Henry Carey, Pottsville, June 11, 1855 (Carey-Gardiner Papers, HSP).

130

It is singular that all the recent explosions should be so clearly traced to carelessness on the part of the miner. But we knew them—even the best miners to be reckless and fearless, and we are only surprised that explosions still more terrible in nature do not often occur.[25]

The miners, on the other hand, blamed the mine's ventilation and instituted a slowdown. Explosions in the St. Clair Shaft thereafter seem to have occurred with some frequency but the record that survives today is spotty because the local newspaper, the Pottsville *Miners' Journal*, did not report some of them, and the private correspondence of the principals did not mention others. There was for instance, an explosion in November 1859 that burned three men and two boys, "some of them severely"; as usual, "the incautious exposure of a naked lamp by a miner" was stated to be the cause.

The most detailed data are provided from 1870 on to the close of the shaft in 1874. During these five years (including the last year, when there was little activity) there were no less than fourteen recorded explosions of fire-damp, causing eight deaths immediately or shortly after and seventeen severe burns (some of which may have eventually proved to be fatal). If one takes an average of two deaths per year from explosions of fire-damp and projects them back through the twenty years of the shaft's operation, the mine must have taken about forty lives from explosions of fire-damp alone.

After McGinness's failure, a succession of operators ran the St. Clair Shaft; the first two of these forfeited their leases because they could not, or would not, bear the cost of paying rent and repairs for a damaged colliery. The first successor, the firm of Kirk and Baum, had barely gotten under way in 1856 when the pump house burned and collapsed into the shaft, destroying the pump itself in the

[25] *Miners' Journal*, October 6, 1855.

131

process; the shaft flooded and was out of operation for a year. The shaft was back in production in 1858, 1859, and 1860, but in 1861 the shaft, beset with pump problems, flooded again. Kirk and Baum, unwilling to pay the expense of repairs and drainage, sold out. The new lessees, a Boston corporation, were unable to start production again until 1864, but in July 1865 the breaker caught fire and the blaze destroyed everything above ground, including the pumps. The New Boston Coal Company withdrew, leaving Henry Carey to spend $26,000 replacing the pumps and draining the mine. Again, in 1871 the shaft was forced to close for a month and a half because an explosion, fire, and consequent flooding of a neighboring mine—the famous Wadesville Shaft—threatened to drown the St. Clair Shaft too. All in all, in its twenty years, the supposedly inexhaustible St. Clair Shaft, which McGinness had expected to yield one thousand tons of coal per day, up to 20,000,000 tons from the 480-acre St. Clair Tract alone, shipped little more than 50,000 tons before it closed forever.[26]

The clearest example of McGinness's imprudence, however, was his failure to construct any air hole at all in a new slope he sank in 1859. He had borrowed money from Henry Carey, and, eager to pay off his debt, he drove the slope, as he said, "as fast as possible." By the end of June 1860 he had reached a depth of 275 yards. On the second of July—before sinking the air hole—he started the gangway east. Two weeks later, on the eighteenth of July, at 2:00 A.M., an explosion of methane gas blew the timbers out of the slope and tore the roof off the engine house.

---

[26] The combined account of the "slope and shaft colliery" at St. Clair has been drawn from a number of sources, principally the following: the letters of Andrew Russell and Enoch McGinness to Henry Carey (Carey-Gardiner Coll., HSP); the Mill Creek and Mine Hill Railroad MSS (Reading Railroad Collection, EMHL); testimony in the trial of Winterstein vs. H. C. Carey et al., 1871 (SCHS); Bowen 1855; and, for the final years, the *Reports of the Mine Inspectors*, 1870-1875.

No one, apparently, was injured or killed. "I am sorry for the accident," he wrote to Carey, "but it was one of the unaccountable things, to happen when there was no work [i.e., no breasts] opened." (It was McGinness who had explained to Carey the same mistake made by Parvin!) On instructions from Carey, he let out contracts at the end of July to repair the works "and at the same time I will have an air hole driven, which will cost about $400." (The $400 was a small fraction of the cost of opening the mine.) By the end of September, however, the air hole was not yet driven, and McGinness, unable to pay the rent, was forced to give up the lease to another miner. In despair, he complained to Carey, "I sometimes think it would be as well to go west and seek for gold at once. I am not the right temperament to stand it."[27]

ECONOMIC REASONS FOR FAILURE:
VENTILATION AS A CURRENT EXPENSE

In both Parvin's and McGinness's cases, a shortage of cash to pay workmen and suppliers prompted the postponement of completing the basic ventilation system. Both men hoped to pay for the completion of the air shafts and furnaces—as well as later extensions of the system—out of income generated by mining and selling coal. The ventilation system, in other words, was a current expense item, to be paid for out of cash receipts, rather than a capital investment. Even when the system was completed, it was, because of continuing financial pressure, apt to be a poor job, constructed as cheaply as possible.

Yet these regressive practices, carried out despite general knowledge of better methods (British immigrants and British handbooks were always present), characterized the leading edge of the American industrialization. In the course

[27] Historical Society of Pennsylvania (HSP), Carey-Gardiner Collection, E. McGinness to Henry Carey, 20 June 1860, 18 July 1860, 21 July 1860, 15 Sept. 1860.

of the Industrial Revolution in the United States the anthracite industry for a time played a major role. It provided the new source of cheap energy, replacing charcoal (and obviating the need for coke), that made possible the great increase in iron and steel production along the east coast in the second half of the nineteenth century.[28] Hard coal also replaced wood as the major fuel in urban home heating and in steam engines that drove ships and locomotives. From the very beginning, however, the industry, particularly in Schuylkill County where St. Clair is located, was beset by economic problems, which manifested themselves in an incessant series of bankruptcies of mining companies. These early difficulties have been variously attributed to an excessive and inappropriately prolonged attachment among coal mine operators to the social values of rugged individualism; to the perennial instability of market prices and labor and transportation costs; to steeply climbing capital costs in operating deep mines; to the prolonged struggles between the operators and the unions; and to the monopolistic pressures exerted by canals and railroads.[29] In recent years, unable to compete with oil and natural gas, deep mining has practically ceased, and strip mining produces only a tenth of what was shipped out a hundred years ago.

The organization of production in Schuylkill County in the 1850s involved four separate groups: the land owners, the colliery operators, the transportation companies, and the colliery employees. The land owners leased the rights to mine coal on their land to operators for varying terms, generally between five and fifteen years, in return for a royalty of thirty to forty cents a ton for every ton shipped to market. A minimum annual tonnage was specified, generally thirty to fifty thousand tons per year, and royalty

[28] See Chendler 1972.
[29] The best study of the economic problems of Schuylkill County anthracite in the early period is Yearley 1961. For the northern counties, see Campbell 1978.

134

on that amount had to be paid whether any coal was actually mined or not. The leases were carefully drawn by attorneys and specified the vein of coal to be mined, the boundaries of the allotted tract, and the use of up-to-date methods of mining. The owner, who lived elsewhere, hired a local agent (often a kinsman) to collect the rents, inspect the mines, and generally look after his interests.

The operator was, at the beginning of the period, expected to raise the capital necessary to open the mine, build the breaker, and install the pumps and other equipment. This, however, was apt to involve capital on the order of $100,000, which had to be borrowed or raised by taking partners. Men with money bought the coal lands directly in preference to investing in troublesome mining operations themselves. As time went on, however, the owners began to assume the capital costs of opening and equipping the collieries. But the operator continued to carry out the necessary geological explorations, design the mine, specify the equipment, and hire, pay, and fire the working employees of the colliery. The colliery operator thus, to varying extents, owned the improvements and capital equipment. The "operator" could be an individual, a partnership, or even a corporation, and some operators ran several mines simultaneously under leases from more than one land owner.

The transportation companies were the railroad and canal corporations. A small lateral railroad took coal from the breaker, where it was dumped directly into cars from overhead chutes, weighed it (on this weight the land owner's royalty was based), and carried it to a distribution center (in this region, Port Carbon) where the cars were either transferred directly to the tracks and locomotives of the Philadelphia and Reading Railroad or re-loaded onto barges at the canal landing of the Schuylkill Navigation Company. The Reading Railroad charged between $1.25 and $1.50 per ton, and the canal about 60¢ per ton to carry the coal to Port Richmond in North Philadelphia. There

135

21. Coal Depot, Port Carbon in Schuylkill County (1877)

it was loaded into the waiting bins of coal merchants and onto ships for other coastal markets. The canal was cheaper, but it was slower and operated only half the year.[30]

Colliery employees, who numbered between 100 and 300 men and boys, were loosely organized by status and occupation. The superintendent was paid an annual salary and was ordinarily expected to be at the site every day. Below him, the employees were divided in roughly equal numbers between inside workers and outside workers headed, respectively, by an inside boss, and sometimes by additional fire bosses, and an outside boss. Outside, above

[30] The Pottsville *Miners' Journal*, a weekly newspaper published by Benjamin Bannan—a coal operator, manufacturer and retailer of mine equipment, and economic philosopher in the following of Henry C. Carey—provided weekly and annual statistics on coal prices, carrying charges, and other detailed information on the coal trade. The writer used the microfilm copy at Eleutherian Mills Historical Library.

136

ground, senior mechanics were responsible for major components of the system: the breaker, the engines, the hoisting machinery, the fan (from the 1860s on), and the machine and carpentry shops. The breaker used a large number of young boys and old and crippled men to pick out rock and other waste from the coal. Inside, there were also boys who opened and closed the doors that controlled ventilation and drove the mules that pulled the wagons between the foot of the shaft and the face of the coal. The heart of the entire operation was the team of miner and laborer who actually mined the coal. Miners were sometimes paid by the ton; these so-called contract miners in turn hired their own laborers and paid for their own lamps, oil, drills, gunpowder, and fuses. As time went on, however, miners and their laborers were increasingly paid by the day, perhaps because the variability of conditions in the mine, and the varying need for "dead-work," required such a frequent renegotiation of the tonnage rates that miners were willing to accept a somewhat lower, but dependable wage. The contract miner, when he was working a good seam, sometimes could earn as much as $25.00 a week; but if he was required to spend much time timbering or climbing up long, steep man-ways to reach the coal face, his income might drop precipitately. Thus, while contract miners *could* make very good money, they were likely to average somewhere around $10.00 to $12.00 per week, and the laborers somewhat less; miners by the day earned even less than this, about $1.50 per day in the 1860s.

The central problem faced by the colliery operator was the relationship between profit, market price, and the cost of production. The operator's gross income was, for practical purposes, the selling price of coal per ton in Philadelphia multiplied by the number of tons sold. The mines in and about St. Clair were considered to be capable of producing about 100,000 tons per year. It was bought by the wholesale merchants of Philadelphia for prices that

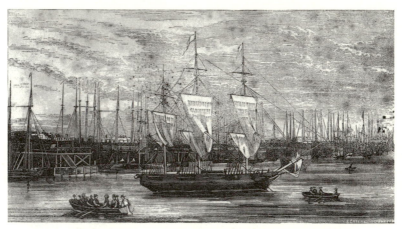

22. Coal Depot, Port Richmond in Philadelphia (1852)

averaged about $3.25 per ton (at retail, the same coal cost householders $4.25 per ton), although there was considerable seasonal variation and longer term variations as a result of general business conditions. The Reading Railroad charged operators about $1.30 per ton, and the Mill Creek and Mine Hill Railroad about 15¢ per ton, so rail haulage (necessary in the winter) amounted to about $1.45 per ton. The cost per ton of mining the coal and hauling it along the gangways to the bottom of the shaft was about 90¢ per ton. Royalties paid to the land owners amounted to about 40¢ per ton. The overhead costs of operating the mine (pumps, hoisting, fans, breakers, extending gangways, deepening shafts, and so forth) could not exceed 50¢ per ton just to break even. Nevertheless, estimates of profit per ton in a colliery in a stable financial condition range from 7¢ per ton in 1871—a deliberately pessimistic figure by the Pennsylvania Mining Inspector—to 50¢ per ton in 1855—a wildly enthusiastic figure by Enoch McGinness. In view of the fact that McGinness was bankrupt a year after he made this prediction, the more conservative figure is probably closer to normal expectation. The Mammoth Vein Consolidated Coal Company, which mined some of

138

the veins near and under the town of St. Clair, estimated in 1864 that "in normal times" (i.e., before the Civil War) they expected a profit of 10¢ per ton on about 70,000 tons of coal per year, and theirs was a successful business.[31]

The operators were acutely conscious that their profits were constantly in danger of being reduced by a slump in the price of coal, or a rise in the toll rates charged by the transportation companies, or a need to pay higher wages. Beyond this, obviously, any factor that interfered with the steady production of coal would lessen their total profit, whatever the profit per ton. During a slowdown or stoppage of production, furthermore, the operator still had to pay part of his production costs. The inside workers and the breaker crew could be laid off, but the steam engines, pumps, and ventilation had to be kept going, lest the mines fill up with water and gas. If the cause of the stoppage were an explosion, fire, or flood, the operator would probably have to spend thousands of dollars on new equipment and pumping. These down-time costs amounted, one operator estimated, to about $150 per day in 1870.[32] And all

[31] The *Miners' Journal* provides weekly information on coal prices and canal and Reading Railroad tolls. Tolls for the St. Clair mines are recorded in the manuscript accounts of the Mill Creek and Mine Hill Railroad, which are to be found in the Reading Railroad Manuscript Collection at Eleutherian Mills Historical Library. Royalties are specified in leases (see for example the leases preserved in the St. Clair Tract Papers, in the Carey-Gardiner Collection at the Historical Society of Pennsylvania). The *Reports of the Inspectors of Coal Mines* for 1871, p. 81, gives a breakdown of average costs. A figure of 12¢ profit per ton was provided by Eckley B. Coxe, a successful operator, in the 1880s (cited in Campbell 1978, 118-119); the Mammoth Vein Consolidated Coal Company in its annual report for 1864 (provided by courtesy of Robert Scheer of the Schuylkill County Historical Society), p. 15, estimated its normal profit at "ten cents a ton." McGinness's optimistic estimate was mentioned in a letter to Henry C. Carey, Pottsville, June 2, 1855 (Correspondence of Henry C. Carey, Carey-Gardiner Collection, Historical Society of Pennsylvania).

[32] The $150/day estimate of "down-time" costs was stated by William Kendrick, operator of the St. Clair Shaft, in testimony before the Pennsylvania Senate, March 16, 1871 (*Report of the Committee . . . on the Anthracite Coal Difficulties*, Harrisburg, 1871, p. 85).

the while he had to keep on paying his 30¢ to 40¢ royalty per ton for coal not even mined. And, finally, the elusive profit had to be turned over immediately, by many of the operators, to the creditors from whom they had borrowed to open their mines.

Dependent on a low and variable profit margin on the coal he shipped, and with neither a cash reserve nor insurance on underground works to back him up, the typical operator was financially highly vulnerable to any disaster that stopped production in his mine. Stoppages forced him to suspend payments on his old debts while he sought to borrow new money not only to repair his colliery but even to pay for routine maintenance expenses and rent. Such disaster-caused colliery failures were a common cause of the bankruptcies that plagued the coal trade.

Clearly their recklessness did not reward Parvin and McGinness financially. Indeed, it is generally agreed that few Schuylkill County mine operators over the long run made money; the sole exception in the St. Clair area was the Johns family, who operated the Eagle Colliery half a mile above the town, and who had a reputation for being solid, cautious, prudent managers. Franklin B. Gowen, president of the Philadelphia and Reading Railroad, who had himself failed in an earlier mining venture, and whose vast and floundering Philadelphia and Reading Coal and Iron Company dragged the parent railroad corporation into two bankruptcy proceedings, stated in his usual flamboyant style in 1875: "Three men retired from the business of coal mining with money. . . . one of these died in an insane asylum and another had softening of the brain."[33] Gowen himself committed suicide a few years later.

The enormity of the danger had for years been realized by the underground miners, and their union had protested unavailingly in Harrisburg until 1869, when a mine safety law was passed that applied to Schuylkill County alone.

[33] Yearley 1961, 65.

The need for a state-wide law was accepted a few months later, when a single-shaft colliery caught fire at Avondale, some miles from St. Clair. Sparks from the ventilation furnace ignited the wooden lining of the shaft; the separate compartments burned and collapsed into the shaft, thus destroying the ventilation; and 115 miners were asphyxiated.[34]

IDEOLOGICAL REASONS FOR FAILURE: THE HERO THEME IN SCHUYLKILL COUNTY

How can we account for such lapses in prudent attention to the necessities of mine ventilation as Parvin's fifteen-yard gap and McGinness's failure to construct any airhole at all in his slope on the Diamond Vein? Clearly enough, economic prudence, as exemplified by the Johns family, required adequate ventilation and timbering; and, at least in the Johns case, such prudence was rewarded by huge profits from the uninterrupted sale of coal, even at low profit margins. But we have also seen that, in England as well as America, economic pressures were inimical to best practice by under-capitalized operators. No substantial amount of correspondence from Parvin survives, but we do have several hundred letters from McGinness to his landlord and patron, Henry Carey, and the mixture of motivations that drove him to gamble away the lives of his workers and his own fortune are visible. He complained constantly of being financially pinched; yet at the

---

[34] The state of art and knowledge in deep-mining technology is described at length in a variety of contemporary publications that were available to the Schuylkill County operators throughout the period under consideration, such as the *Franklin Journal* and the *Miners' Journal*, both of which carried numerous articles on the subject of ventilation and gas explosions, various British treatises, and a standard text written by Daddow and Bannan in 1866. An account of the Avondale disaster is given in Pinkowski 1963. The Acts of 1869 and 1870 are printed in the *Laws of the General Assembly of the State of Pennsylvania*, sessions of 1869 and 1870 (Harrisburg 1869 and 1870).

same time, he always believed he was only days away from wealth. The royalty payment, the men's wages, the notes were always coming due; with his last dollar, gleaned from pawning his wife's silver tea service, he was straining to break through into marketable coal.

But there was more to McGinness's risk-taking behavior than a gambler's impulse. McGinness wanted to be admired as a daring innovator, a formulator of new geological theories, a pioneer who led the way for others, a leader whose sacrifice made possible the wealth of others. He wrote of the coal trade's need for "daring men" who were devoted to the advancement of the coal region; he deprecated those like William Johns, who merely wanted to make money by digging coal conservatively and selling it at a profit. He wanted to be a hero.

For—to return to the theme with which we began— Schuylkill County was suffused with respect for the economic hero. To an extent, the ambience of industrial heroism was expressed in the region's classrooms and lyceums and libraries, where the views of Samuel Smiles and other myth-makers were declared. But it was communicated more directly outside the classroom, in the community of adults. Here was a world of more public symbols, conveyed in public ceremonies, entertainments, books and periodical literature, in which deliberate efforts were made to articulate a merging of the heroic image of military and economic or industrial patriotism.

The most conspicuous symbol in the St. Clair region was the Henry Clay monument. The Henry Clay monument was, and is, a colossal iron statue of the American statesman, set upon a cast-iron Doric column, that stands 66 feet in total height from the base to the top of the head, and twice that above the main street of the town. It is visible from the north for miles, and still dominates the town below it. Henry Clay was a hero to the industrialists of the coal region for one reason: his steady advocacy, in the face of southern opposition, of a protective tariff that

142

Henry Clay Monument,
Pottsville, Pa.

23. Henry Clay Statue at Pottsville

143

nourished the rapidly growing Pennsylvania iron industry and thus the coal industry that supplied it with fuel. Just as in cotton manufacturing towns there were mills named after him, so in the anthracite district there was a town named for his estate, Ashland, and a mine named for the man himself. When Clay died in 1852, John Bannan, ever active in civic and educational affairs, and Samuel Silliman, land owner and mine operator of the borough, proposed erecting a monument in his memory. The construction, raising, and dedication were arranged by a variety of committees in which the men in the coal trade were prominent, and the monument was actually completed and dedicated on the 4th of July. Invitations to participate were distributed nationally. The grand procession was composed of thirty-three units, starting with six military units (including two generals and two regiments of infantry), extending through a long list of bands, occupational categories, and benevolent associations, and ending with "citizens mounted." The public meeting was replete with distinguished military and political figures, and an orator delivered a special address "teeming with great thoughts." It was a drunken and joyous celebration of the union of military and economic patriotism. The resemblance of the monument to Nelson's Column in Trafalgar Square is unmistakable.[35]

This theme—the union of the military and industrial interests of the region—is also the motif of a curious piece of "anthracite drama" composed by Colonel John Macomb Wetherill—mining agent for the Wetherill family— and published by John Bannan in 1852, after being successfully presented to the social circle of the region. Entitled *Mars in Mahantango: A Play in Five Acts*, it is a comedy of manners whose plot concerns an election for Brigadier General of the local militia regiment. The hero,

[35] The account of the Henry Clay Monument is drawn from Elssler 1910.

144

Captain Maxwell, is a ladies' man who has fallen in love with beautiful Caradori, daughter of Conglomerate, a wealthy and unscrupulous mine operator, founder of the boom town of New Babylon. Conglomerate goes bankrupt and departs for California; Maxwell wins the election and the heart of Caradori (he saves her from a rattlesnake while on a country walk), and they marry. The cast of characters includes social types familiar in the district—Tangent, an engineer; Gunsenhauser, the Pennsylvania-German tavern keeper and father of the buxom and compliant Kunigunde (who constantly astonishes her swains by quoting at length from French authors); Blackstone, a politician and lawyer; and several others. Wetherill was, in fact, writing about local types and issues for a local audience. He himself was enthusiastically interested in two things: the administration of his family's and his own coal properties, and a career in the militia, in which he held the rank of Colonel by 1855. He was no mere sunshine patriot—during the Civil War, he was involved in major battles. He notes, and pokes fun at, the grandiose dreams of engineers like Tangent whose projected bridge over Mill Creek, a part of Conglomerate's new scheme, was to be

> A structure, such as modern Art
> Hath never known; eclipsing even that
> The Roman despot gave unto the Goths.

Conglomerate aimed to "be another Nebuchadnezzar" but was roundly criticized for risking the welfare of other investors in his "stupendous schemes" by "sinful recklessness." Conspicuously absent from this example of anthracite drama are two kinds of people: actual miners and land owners.[36]

During the tense winter days preceding the first sum-

---

[36] The only printed copy of Wetherill's play is located at the Historical Society of Schuykill County in Pottsville.

145

mer of the War Between the States, the same conjunction of Mars and anthracite was conspicuously present. An "Inauguration of an American Flag" took place upon the steps of the county court house in Pottsville on February 22, 1861 (Washington's birthday), followed by a procession and speeches. The procession was as usual led by military units, now including the "Wetherill Rifles of St. Clair" and the "Marion Rifles of Port Carbon," headed by Captain Allison (an engineer, also of St. Clair). John Bannan, of course, was the organizer; he was now a colonel himself. The speech, supporting the union, was richly embroidered with allusions to "a noble history" and to "the illustrious deeds" of the hearers' ancestors.[37]

Promotional literature extolling the contributions to national greatness made by men of industry was common not only in the pages of the *Miners' Journal*. On a national scale, it was voiced perhaps most effectively by J. Leander Bishop in his classic *History of American Manufactures*.

24. View of Pottsville with Clay Statue (1852)

[37] The pamphlet celebrating the inauguration is in the Historical Society of Pennsylvania, Schuylkill County Pamphlets.

146

He took as his goal to rescue from oblivion the "true history" of the men whose industrial innovations and enterprises built up "the fair fabric of our national civilization." He sought not merely to give an account of their material contributions but also an appreciation of their lives:

> what were their everyday pursuits, how they lived and supported their families, and shaped the character or directed the channels of American labor, as well as to know their lineage and connections, for whom they voted, and how they fought. Unfortunately, history has been too little cognizant of anything but the public acts or words of the world's benefactors; while often the most instructive examples of their struggles and triumphs, the heroism of their daily life, is consigned to a narrower influence.[38]

Such commentaries could be cited in considerable numbers, and the work of myth-making does not cease. A brief biography of Asa Packer, the patriarch of the coal industry in the Lehigh Valley and the founder of Lehigh University, was published by The Newcomen Society in North America in 1938. It is entitled *Asa Packer 1805-1879: Captain of Industry; Educator; Citizen*, and the author, Milton C. Stuart (a graduate engineer from the University of Pennsylvania), delineates the hero in a style reminiscent of Samuel Smiles. He quotes approvingly from a eulogy delivered at the time of Asa Packer's death that once again praises the union of military and civilian (even feminine) virtues.

> He was both gentle and inflexible, persuasive and commanding, in his sensibilities refined and delicate as a woman, and in his intellect and resolve clear and strong as a military leader; pliant as the limbs of a

[38] Bishop 1864, 1: 9.

147

tree waving to the touch of the breeze, and sturdy as the trunk which defies the tempest.

And the work concludes with the lines from Ecclesiastes XLIV: 1. "Let us now praise famous men. . . ."[39] What evidence is there that this sort of rhetoric was actually internalized in the head of Enoch McGinness? As we have seen, in the early 1850s he interested Henry Carey in a theory that the Mammoth Vein could be reached, under the town of St. Clair, by a vertical shaft to a moderate depth of about five hundred feet. This theory was greeted derisively when it was first expounded (it is satirized in Wetherill's play as the "Magnesian theory" that lured Conglomerate into bankruptcy). But McGinness succeeded, and his accomplishment earned him immediate honor (but not wealth). On the evening of October 13, 1854, at the Mount Carbon Hotel in Pottsville, he was praised at a testimonial dinner attended by some of the most important men in the coal trade (including Colonel Wetherill, now reconciled to the Magnesian theory). The toasts must have amply filled his cup of pride, for he heard himself lauded as a man of "genius, energy, and courage" who had battled obstacles of scientific prejudice, personal poverty, and hard rock in making the great, the fabled

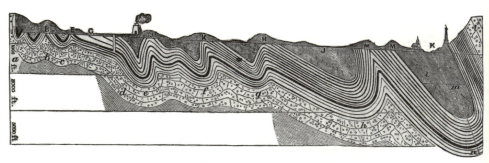

25. Diagram of Coal Measures with Clay Statue (1866)

[39] Stuart 1938, 19.

148

Mammoth Vein accessible to man, in having proved the practicality of deep-shaft mining, and for having shown that the anthracite mines of Pennsylvania were inexhaustible. McGinness's former rival, Colonel Wetherill, was the most fulsome of all, recognizing McGinness as one of those who, in language anticipatory of Samuel Smiles, had contributed most importantly to "the rescue of the country from the domination of the beasts of the forest, to place it under the rules of an enlightened life." He concluded by raising his glass to "Enoch McGinness, the Columbus of Schuylkill County!" The honorable F. W. Hughes, Attorney General of Pennsylvania, declared that the "McGinness theory of the Schuylkill County coal formation" had shown that the county possessed "a supply of fuel for all the purposes of the thronging myriads that in future generations will inhabit this continent." In response, McGinness (compared in print to Marc Antony by the author of the memorial of the occasion, and characterized as "uniting delicacy of feeling with the stamina of a giant") delivered the classic hero's acknowledgment as he received the cup of victory, conveying a becoming modesty, a due awareness of the help of others, an intention only to pay his debts and support his family, and his satisfaction at having "been instrumental in benefitting others." He averred that he expected to market "one thousand tons of Coal per day from the St. Clair shaft." (But at the date of the banquet he had actually brought to the surface only one hundred tons, from which he might expect to gain a profit-over-cost of about five dollars.)[40]

Conclusion

Let us say, for the sake of argument, that there were others like Enoch McGinness in the anthracite district—and, indeed, in other coal regions and in other industries—who

[40] Bowen 1855. Copy located at Eleutherian Mills Historical Library.

wanted both wealth and honor, and to that end were willing to wager their own and others' lives and fortunes on the last fifteen yards—i.e., on the chance of avoiding a gas explosion in a mine with an incomplete ventilation system. The logic is compelling. If the gamble succeeds, the operator saves money, gets an earlier profit, proceeds to complete the ventilation system, and earns praise for running the risk of bankruptcy on behalf of his creditors and the community. The risk is highly public: all the miners, their families, and the local mining engineers know about it. If he fails—as, sooner or later, he does—he does in fact go into bankruptcy, but he is still the public's benefactor for having opened the mine. It would seem, then, that the prospect of honor, in the ambience of the industrialist-as-hero theme, provided a margin of motivation that made it possible for the Schuylkill County mines to provide Pennsylvania's iron industry with anthracite at a price below the cost of production, the difference being made up by dead miners and the creditors of bankrupt operators.

For the development of an adequate safety technology, however, the function of the industrial hero myth was generally negative. It was the ventilation system that was regularly disregarded by operators who did not bother to make use even of the technical knowledge readily available to them in English and American treatises on coal mining methods. In McGinness's St. Clair Shaft, the coal was hauled to the surface in a newly invented elevator; but the old-fashioned ventilation system McGinness designed never worked. It is hardly the world of heroic inventors that Bacon foresaw in *New Atlantis*. To the contrary, in certain of its domains—the high-pressure steam engine and the coal mine, at least—the Industrial Revolution would appear to be a game played by technological gamblers who liked to bet their own and others' lives and money against disaster, in the hope in part of the noneconomic reward of the industrial hero's accolade. Do we see here the prelude to environmental disasters of the twentieth century?

150

CONCLUSION

IN THE Introduction, we observed that Protestantism generally, let alone Puritanism specifically, while favorable to the technological innovations of the early Industrial Revolution, was hardly a necessary pre-condition. It was state policy in Catholic continental countries and in Anglican England to encourage technological and scientific innovation. Searching to define the social circumstances that nourished the new technology, we turned to three case studies: the invention of the steam engine in England in the seventeenth century and in the eighteenth, up to the succesful machine marketed by Newcomen and Savery; the substitution of coke for charcoal in the smelting, founding, and refining of iron in eighteenth-century England; and the development in England and America of procedures for reducing the danger of fire-damp explosions in coal mines by improved techniques of ventilation in the eighteenth and nineteenth centuries. The latter case also provides something of a control situation in which in some mines not only did innovation not occur but well-known best practice was not followed.

What conclusions can we draw, however tentatively, from these case studies? It would appear that the successful pursuit of the courses of innovation considered here involved three very specific institutional characteristics, and two rather general ones. The specific features are:

(1) an institution that survives over two or more generations, permitting an "in-house" transfer and concentration of technical information in a small paradigmatic community;

151

(2) an institution that controls or has access to redundant resources in money, land, physical plant, and skilled labor;

(3) an institution that regards the support of innovation at its own expense as part of investment in best practice.

These features could be (but were of course not necessarily) found in two types of institutions: an office of government (as in the case of the Royal Office of Ordnance) and an extended family, particularly a cognatic descent group, upon which an estate was entailed or with which a business enterprise was otherwise associated. Two other institutional forms also were involved but played a subsidiary role: the partnership, which tended to be a short-lived association; and the descendants of the old guilds—the English companies of merchants and later trade associations—which were more apt to restrict innovation than to encourage it, out of fear of allowing unequal advantage to their more ingenious members. In principle the limited liability corporation could also satisfy these conditions, but, although it was significant in other areas, it did not become important as a context of technological innovation until well into the nineteenth century. Two other institutions, the church and the university, might also have served, being both transgenerational and capital redundant. Indeed, in the medieval period, some monasteries had been centers of technological innovation; and Bacon, as we have seen, correctly foresaw the role of the university and the R & D institute in nourishing new technology. But the monastic orders had been abolished in England, and her universities were still, despite Bacon's urging, centers of scholasticism. Nor was the individual capitalist in a competitive market situation likely to innovate, both because of economic pressure (which pressed him to postpone investment in best practice in favor of immediate cash income) and, in the later years, in cases of industrial safety, because non-economic rewards (the

152

"industrial hero" theme) may have further diminished his interest in being altogether secure.

The two general social conditions would seem to be:

(1) a general ambience of encouragement of technological as distinct from scientific innovation as a contribution to social progress;

(2) a porous social structure that permitted lower- and middle-class inventive artisans and innovative entrepreneurs, including religious non-conformists and recent immigrants as well as natives, to acquire wealth, property, and social position, and that also allowed upper-class gentlemen and nobles to associate with these members of the lower orders in a common concern with the solution of technological problems.

The first of these aspects of the technological ethos would seem to have characterized England and the United States earlier than France, Germany, and other Western countries, where—as in the case of seventeenth- and eighteenth-century France—science was far more respectable than technology, which at best was subsumed under science as a mere application of known scientific principles. (Actually, of course, the development of technology forces scientists to frame questions that have not hitherto been asked—witness the relation between the steam engine and the science of thermodynamics.) The second aspect of the technological ethos also emerged earlier in England and the United States than in other Western countries and made possible a rapid communication and collaboration of persons with diverse skills, experience, and intuitions that a more rigid social structure—again like that of France—made difficult or impossible.

The historical process by which these specific cultural and institutional features, none of them peculiar to England, came together there and in her colonies sooner than in other places is not our business to explore here. We have attempted to examine the specific institutional set-

153

tings in which important events of irreversible technological change involving steam, iron, and coal occurred in
the early Industrial Revolution, and to infer from these
cases certain of the social and cultural pre-conditions of
the initiation of that great event whose end we still cannot
see.

BIBLIOGRAPHY

ABBREVIATIONS IN FOOTNOTES

DNB: Dictionary of National Biography
EMHL: Eleutherian Mills Historical Library (Wilmington, Del.)
HSP: Historical Society of Pennsylvania (Philadelphia, Pa.)
NBG: Nouvelle Biographie Generale
SCHS: Schuylkill County Historical Society (Pottsville, Pa.)

Addis, John P. *The Crawshay Dynasty: A Study in Industrial Organization and Development, 1765-1867.* Cardiff: University of Wales, 1957.

Agricola, Georgius. *De Re Metallica* (trans. and ed. by Herbert C. and Lou N. Hoover). New York: Dover, 1950 (1556).

Ashton, Thomas S. *Iron and Steel in the Industrial Revolution.* Manchester: Manchester University Press, 1924.

———. "The Discoveries of the Darbys of Coalbrookdale," *Transactions of the Newcomen Society* 5 (1926): 9-14.

———. *Iron and Steel in the Industrial Revolution.* Manchester: Manchester University Press, 1963. 3rd edition.

Atkinson, A. A. *A Key to Mine Ventilation.* Scranton: The Colliery Engineer Co., 1892.

Aubrey, John. *The Natural History and Antiquities of the County of Surrey.* 1719. Reprinted with an intro. by J. L. Nevinson. Dorking: Kohler and Coombes, 1975.

Bacon, Francis. *The Works of Francis Bacon*, ed. by James Spedding, Robert Ellis, and Douglas Heath, Vols. 1 and 2. Boston: Houghton, Mifflin, n.d.

155

Bagot, Alan. *The Principles and Practice of Colliery Ventilation*. Birmingham: Osborne, 1879.

Beard, James T. *Mine Gases and Ventilation*. New York: McGraw-Hill, 1920.

Birch, A. "Carron Company, 1784-1822, the Profits of Industry During the Industrial Revolution," *Explorations in Entrepreneurial History* 8 (Dec. 1955): no. 2.

Birch, Thomas. *A History of the Royal Society*. London: Printed for A. Millar, 1756-57. Volume 1.

Biringuccio, Vannoccio. *The Pirotechnia*. New York: Basic Books, 1942 (1540; trans. into English).

Bishop, Leander J. *A History of American Manufactures*. Philadelphia: Young, 1864 (2 vols.).

Bolton, Henry C. *The Follies of Science at the Court of Rudolph II (1576-1612)*. Milwaukee: Pharmaceutical Review, 1904.

Bowen, Eli. *The McGinness Theory of the Schuylkill Coal Formation*. Pottsville: Bannan, 1855.

Buckler, John Chessell. *An Historical and Descriptive Account of the Royal Palace at Eltham*. London: Nichols, 1828.

Campbell, R. H. *Carron Company*. Edinburgh: Oliver and Boyd, 1961.

Campbell, Stuart A. "Business and Anthracite: Aspects of Change in Late Nineteenth Century." Unpublished Doctoral Dissertation, University of Delaware, 1978.

Carey, Henry C. Correspondence (manuscript). Carey-Gardiner Collection, Historical Society of Pennsylvania.

———. *Letter to . . . the Stockholders of the St. Clair Coal Company of Boston*. Philadelphia, 1865.

Chance, H. M. *Report on the Mining Methods and Appliances used in the Anthracite Coal Fields*. Harrisburg: 2nd Geological Survey, 1883.

Cipolla, Carlo M. *Guns, Sails and Empires: Technological Innovation and the Early Phases of European Expansion, 1400-1700*. New York: Pantheon Books, 1965.

156

Colie, Rosalie. "Cornelius Drebbel and Salomon de Caus: Two Jacobean Models for Salomon's House," *Huntington Library Quarterly* 18 (1955): 245-260.

Court, W.H.B. *The Rise of the Midland Industries 1600-1838.* London: Oxford University Press, 1938.

Cunningham, W. *Alien Immigrants to England.* London: Sonnenschein, 1897.

Daddow, Samuel H., and Benjamin Bannan. *Coal, Iron, and Oil: or, the Practical American Miner.* Pottsville: Bannan, 1866.

Darby, Abiah. "Extracts from the Diary of Abiah Darby," *Journal of the Friends Historical Society* 10 (1913): 79-92.

Davy, Humphry. *On the Safety Lamp for Coal Mines.* London: Hunter, 1818.

Dickinson, H. W. *John Wilkinson, Ironmaster.* Ulverston: Hume Kitchin, 1914.

———. *A Short History of the Steam Engine.* Cambridge: Babcock and Wilcox, 1938.

———. *Sir Samuel Morland, Diplomat and Inventor, 1625-1695.* Cambridge: Newcomen Society/W. Heffer and Sons, 1970.

Dircks, Henry. *The Life, Times, and Scientific Labours of the Second Marquis of Worcester.* London: Quaritch, 1865.

Doorman, G. "The Marquis of Worcester and Caspar Calthoff," *Transactions of the Newcomen Society* 26 (1947-49): 269-271.

Dudley, Dud. *Metallum Martis: or, Iron made with Pit-coale, Sea-coale, &c.* London, 1665.

Elssler, Ermina. "The History of the Henry Clay Monument," *Publications of the Historical Society of Schuylkill County* 2 (1910): 405-417.

Evans, R.J.W. *Rudolf II and His World: A Study in Intellectual History 1576-1612.* Oxford: Oxford University Press, 1973.

Evelyn, John. *Diary and Correspondence*, ed. by William Bray. London: George Bell, 1889. 4 vols.

Ferguson, Eugene S. "The Origins of the Steam Engine," *Scientific American*, 210 (1964): 98-107.

Ffoulkes, C. J. *The Gunfounders of England*. Cambridge: Cambridge University Press, 1937.

Flinn, M., ed. *Law Book of the Crowley Iron Works*. London: Quaritch, 1957. (Publications of the Surtees Society, no. 167.)

———. "The Growth of the English Iron Industry, 1660-1760," *Economic History Review* 11 (1958): 144-153.

———. "The Lloyds in the Early English Iron Industry," *Business History* 2 (1959): 21-31.

———. *Men of Iron: The Crowleys in the Early Iron Industry*. Edinburgh: Edinburgh University Press, 1962.

Fox, Robin. *Kinship and Marriage*. Harmondsworth: Penguin, 1967.

Galilei, Galileo. *Dialogues concerning the Two New Sciences* (trans. by Henry Crew and Alfonso de Salvio). Evanston: Northwestern University Press, 1939 (1638).

Gerland, Ernst, ed. *Leibnizens und Huygens' Briefwechsel mit Papin*. Wiesbaden: Sändig, 1966 (1881).

Haas, J. M. "Work and Authority in the British State Shipyards from the Seventeenth Century to 1870," *Proceedings of the American Philosophical Society* 124 (1980): 419-428.

Hall, J. W. "Notes on Coalbrookdale, and the Darbys," *Transactions of the Newcomen Society* 5 (1926): 1-8.

Harris, L. E. *The Two Netherlanders: Humphrey Bradley and Cornelius Drebbel*. Leiden: E. J. Brill, 1961.

Hedley, John. *A Practical Treatise on the Working and Ventilation of Coal Mines: with suggestions for improvements in mining*. London: John Weale, 1851.

Hodgen, Margaret T. *Change and History: A Study of the Dated Distributions of Technological Innovations in*

England. New York: Wenner-Gren Foundation, 1952. (Viking Fund Publications, no. 18.)

Hogg, O.F.G. "The Development of Engineering at the Royal Arsenal," *Transactions of the Newcomen Society* 32 (1959-60): 29-42.

————. *The Royal Arsenal: Its Background, Origin, and Subsequent History*. London: Oxford University Press, 1963.

Hollister-Short, G. "Leads and Lags in Late Seventeenth Century English Technology," in *History of Technology, 1976*, ed. by A. Rupert Hall and Norman Smith. London: Mansell, 1976: 159-183.

————. "The Vocabulary of Technology," in *History of Technology 1977*, ed. by A. Rupert Hall and Norman Smith. London: Mansell, 1977: 125-155.

Hunt, B. C. *The Development of the Business Corporation in England, 1800-1867*. Cambridge: Harvard University Press, 1936.

Jenkins, Rhys. "A Sketch of the Industrial History of the Coalbrookdale District," *Transactions of the Newcomen Society* 4 (1923-24): 102-107.

————. "Iron-making in the Forest of Dean," *Transactions of the Newcomen Society* 6 (1926-27): 42-65.

————. *Links in the History of Engineering and Technology from Tudor Times*. Freeport: Books for Libraries, 1971 (first published 1936).

Johnson, B.L.C. "The Foley Partnerships: the Iron Industry at the End of the Charcoal Era," *Economic History Review* 2nd ser., 4 (1951): 322-340.

————. "The Midland Iron Industry in the Early Eighteenth Century: the Background to the First Successful Use of Coke in Iron Smelting," *Business History* 2 (1960): 67-74.

Lane, Frederic Chapin. *Venetian Ships and Shipbuilders of the Renaissance*. Baltimore: Johns Hopkins, 1934.

Laws of the General Assembly of the State of Pennsyl-

vania. Sessions of 1869 and 1870. Harrisburg, 1869, 1870.

Maddison, John. Engineer's Reports. Schuylkill County Historical Society.

Mammoth Vein Consolidated Coal Company. Annual Report for 1864. Schuylkill County Historical Society.

Mason, Rev. Mr. "Concerning Spelter, Melting Iron with Pit-coal; and on a Burning Well at Broseley," *Philosophical Transactions* 9 (1744-49): 305-306.

Merton, Robert K. *Science, Technology and Society in Seventeenth-Century England.* New York: Harper, 1970.

Mill Creek and Mine Hill Railroad MSS. Reading Railroad Collection. Eleutherian Mills Historical Library.

Moore, Ralph. *The Ventilation of Mines.* Glasgow and London: Hamilton, Adams, 1859.

Moran, Bruce T. "German Prince-Practitioners: Aspects of Courtly Science, Technology, and Procedures in the Renaissance," *Technology and Culture,* 22 (1981): 253-274.

Morland, Samuel. *Elevation des Eaux par toute sorte de machines reduite a la Mesure, au Poids, a la Balence, par le moyen d'un nouveau piston, & corps de pompe, & d'un nouveau movement cyclo-elliptique, au rejettant l'usage de toute forte de maniuelles ordinaires; avec huit problemes de machanique proposez aux plus habiles & aux plus scavans du siecle.* Paris, 1685.

Mulligan, Lotte. "Puritans and English Science: A Critique of Webster," *Isis* 71 (1980): 456-469.

Multhauf, Robert P. "Mine Pumping in Agricola's Time and Later," *Contributions from the Museum of History and Technology, U. S. National Museum Bulletin* 128 (1959): 114-120.

Needham, Joseph. "The Pre-Natal History of the Steam Engine," *Transactions of the Newcomen Society* 35 (1962-63): 3-58.

Nef, John U. *Industry and Government in France and England, 1540-1640.* Philadelphia: American Philosophical Society, 1940. 2 vols.

————. *The Rise of the British Coal Industry.* Hamden: Archon Press, 1966 (1933).

North, Roger. *The Lives of Francis North, Dudley North, and John North.* London: Colburn, 1826.

Paterson, Antoinette M. *Francis Bacon and Socialized Science.* Springfield: Thomas, 1973.

Pinkowski, Edward. "Joseph Battin: Father of the Coal Breaker," *Pensylvania Magazine of History and Biography* 73 (1949): 337-348.

————. *John Siney.* Philadelphia: Sunshine Press, 1963.

Price, W. N. *The English Patents of Monopoly.* Boston: Houghton, Mifflin, 1906.

Pullan, Brian. *Crisis and Change in the Venetian Economy in the Sixteenth and Seventeenth Centuries.* London: Methuen, 1968.

Raistrick, A., and E. Allen. "The South Yorkshire Ironmasters, 1690-1750," *Economic History Review* 9 (1939): 168-185.

Raistrick, A. *Quakers in Science and Industry.* New York: Philosophical Library, 1950.

————. *Dynasty of Iron Founders: The Darbys and Coalbrookdale.* London: Longmans, Green, 1953.

Ramelli, Agostino. *The Various and Ingenious Machines of Agostino Ramelli (1588),* ed. by Eugene S. Ferguson. Baltimore: Johns Hopkins, 1976.

Randall, John. *The Wilkinsons.* Madeley, n.d.

————. *History of Madeley, including Ironbridge, Coalbrookdale, and Coalport.* Madeley, 1880.

Rapp, Richard. *Industry and Economic Decline in Seventeenth Century Venice.* Cambridge: Harvard University Press, 1976.

Rathbone, H. M. *Letters of Richard Reynolds, with a Memoir of His Life by his Granddaughter Hannah Mary Rathbone.* London: Gilpin, 1852.

161

Reading Railroad Collection. Eleutherian Mills Historical Library.

Rees, William. *Industry before the Industrial Revolution, incorporating a study of the Chartered Companies of the Society of Mines Royal and of Mineral and Battery Works*. Cardiff: University of Wales Press, 1968.

Report of the Committee . . . on the Anthracite Coal Difficulties. Harrisburg, 1871.

Reports of the Inspectors of Coal Mines, 1869-1875. Harrisburg, Pa.

Rolt, L.T.C. *Thomas Newcomen: The Prehistory of the Steam Engine*. London: MacDonald, 1963.

Rossi, Paolo. *Francis Bacon: from Magic to Science*. Chicago: University of Chicago Press, 1968.

Rovinson, John. *A Treatise of Metallica*. 1613.

Royal Society of London. *The Philosophical Transactions of the Royal Society of London, from their Commencement, in 1665 to the year 1800*. Abridged with notes and biog. by Charles Hutton, George Shaw, and Richard Pearson. Vol. 1, 1665-1672. London, 1809.

Rye, William Brenchley. *England as Seen by Foreigners in the Days of Elizabeth and James I*. London: John R. Smith, 1865.

St. Clair Tract Papers. Carey-Gardiner Collection. Historical Society of Pennsylvania.

Sarjeant, William A. S. "The Dudley Canal Tunnel and Mines, Worcestershire," *The Mercian Geologist* 1 (1964): 61-66.

Savery, Thomas. *Navigation Improved: Or, the Art of Rowing Ships of all Rates, in Calms. . . .* London: John Moxon, 1698.

———. "An Account of Mr. Tho. Savery's Engine for Raising Water by the Help of Fire," *Philosophical Transactions* (1699): no. 253.

———. *The Miner's Friend; or An Engine to Raise Water by Fire*. London: Crouch, 1702 (reprint 1827).

Schubert, H. R. *History of the British Iron and Steel Industry from c. 450 B.C. to A.D. 1775.* London: Routledge and Kegan Paul, 1957.

Smiles, Samuel. *Self-Help; with Illustrations of Character, Conduct, and Perseverence.* New York: Lovell, n.d. (revised edition; first edition 1857).

———. *The Life of George Stephenson* (1859). New York: Harper & Bros., 1868.

———. *Character.* New York: Harper, 1877.

———. *Duty, with Illustrations of Courage, Patience, and Endurance.* New York: Harper, 1881.

———. *Lives of the Engineers: Selections from Samuel Smiles,* ed. by Thomas P. Hughes. Cambridge: M.I.T. Press, 1966.

Smith, Adam. *An Inquiry into the Nature and Causes of the Wealth of Nations* (1776), ed. by Edwin Cannan. Chicago: University of Chicago Press, 1976.

———. *The Theory of Moral Sentiments; or; an Essay Towards an Analysis of the Principles by Which Men naturally judge concerning the Conduct and Character, first of their Neighbors, and afterwards of themselves.* Glasgow: R. Chapman, 1809.

Smith, George K. *Engineering Notebook.* Schuylkill County Historical Society.

Smith, Merritt Roe. *Harper's Ferry Armory and the New Technology: The Challenge of Change.* Ithaca: Cornell University Press, 1977.

Straker, E. *Wealden Iron.* Newton Abbott: David and Charles, 1969.

Stuart, Milton C. *Asa Packer, 1805-1879; Captain of Industry; Educator; Citizen. . . .* Princeton: Princeton University Press, 1938.

Sturtevant, Simon. *Metallica.* London, 1612.

Surtz, Edward S.J., and J. N. Hexter. *The Complete Works of St. Thomas More.* New Haven: Yale University Press. Vol. 4 (1965): *Utopia.*

Switzer, Stephen. *An Introduction to a General System*

of Hydrostaticks and Hydraulicks, Philosophical and Practical. London, 1724.

Thorpe, W. N. "The Marquis of Worcester and Vauxhall," *Transactions of the Newcomen Society* 13 (1932-33): 75-88.

Tierie, Gerrit. *Cornelis Drebbel (1572-1633).* Amsterdam: Paris, 1932.

Tomlinson, Howard C. "Wealden Gunfounding: An Analysis of Its Demise in the 18th Century," *Economic History Review* 29 (1976): 383-400.

Triewald, Marten. *Short Description of the Atmospheric Engine.* Stockholm, 1734.

Trinder, Barrie. *The Industrial Revolution in Shropshire.* Totowa: Rowman & Littlefield, 1973.

Unwin, George. *Industrial Organization in the Sixteenth and Seventeenth Centuries.* Oxford: Clarendon Press, 1904.

Victoria County History of Kent.

Victoria County History of Surrey.

Walford, Edward. *Old and New London: A Narrative of Its History, Its People, and Its Places.* London: Petter and Galpin, n.d.

Wallace, Anthony F. C. "Paradigmatic Processes in Culture Change," *American Anthropologist* 74 (1972): 467-478.

————. *Rockdale.* New York: W. W. Norton and Company, 1980.

Webster, Charles. *The Great Instauration: Science, Medicine and Reform, 1626-1660.* London: Duckworth, 1975.

Westfall, Richard S. *Never at Rest: A Biography of Isaac Newton.* Cambridge: Cambridge University Press, 1980.

White, Josiah. *Josiah White's History, Given by Himself.* Philadelphia: Lehigh Coal and Navigation Co., 1909.

White, Lynn, Jr. *Medieval Technology and Social Change.* London: Oxford University Press, 1962.

Wilkins, Charles. *The History of Iron, Steel, Tinplate, and . . . Other Trades of Wales.* Merthyr Tydfil: Jos. Williams, 1903.

Winterstein vs. H. C. Carey. Testimony. Schuylkill County Historical Society.

Woodbury, Robert S. *Studies in the History of Machine Tools.* Cambridge: M.I.T. Press, 1972.

Young, W. A. "Works Organization in the Seventeenth Century: Some Account of Ambrose and John Crowley," *Transactions of the Newcomen Society* 4 (1923-24): 73-93.

Zonca, Vittorio. *Novo teatro di machine et edificii. . . .* Padova: Bertelli, 1607.

INDEX

Galileo, xii; *Discourses on Two New Sciences*, 29
Garbett, Samuel, 98
"Gas as a Ventilator" (Daddow and Bannan), 125
German immigrants in England, 36
German mining methods, 32, 36-37, 118
Glasgow (Scotland), 98
Gobelins manufactory, 23
Gowen, Franklin B., 140
Greenwich (England), 21-23, 27, 35
Guest family, 72, 76, 88
gunpowder engine, 47, 55, 57

Hall, Benjamin, 75
Hall, Mrs. Charlotte, *see* Crawshay, Charlotte
Hampton Court (London), 50
Harrisburg (Pennsylvania), 140
Hartlib, Samuel, 39
Harvey, William, xii
Hawkins, John, 87-88, 96-98
Hawkins, Robert, 99
Heidelberg (Germany), 34
Henry III of Germany, 30
Henry, Prince of Wales, 22, 34
hero, industrial, 103-109, 125, 141-50, 152-53
Hero of Alexandria, 34
Hogge, Richard, 70
Hooke, Robert, 47, 55
Hornblower, Jonathan, 93
Horsehay (Shropshire), 97
House of Commons (British), 14, 41
Hughes, F. W., 149
Huguenots, 5-6
Humfrey, William, 69-70
Huygens, Christian, 47, 55, 57

industrial hero, *see* hero, industrial

Industrial Revolution, xi-xii, 3, 8, 78, 91, 108, 134, 150-51, 154
innovation, conditions of, 151-54; and administrative style, 6-7; and religion, 4-6
Inspector of mines (Pennsylvania), 138
iron, smelting and refining of, *illus.* 92; blast furnace introduced, 67-69, *illus.* 68, 100; bloomery furnaces, 67, 69, 100; hot blast introduced, 98-99; rolling and slitting mills, 69, 72, 74, 100; substitution of anthracite for charcoal in, 122, 134; substitution of coke for charcoal in, 86-88, 90-91, 100, 151
Iron Bridge, 93, *illus.* 95
Italian family, 71

James I of England, 12, 19, 22, 25
Jenkins, Rhys, 60
Johns, William H., 123-25, 142
Johns family, 123-25, 140-41
Johnson, Samuel, 74
Journal of the Franklin Institute, 116, 141

Kalthoff, Caspar, 5, 38, 40-42, 45-47, 50, 52
Kendrick, William, 139
Kent (England), 7, 25-27, 68, 70
Kepler, Johannes, 23
Ketley (Shropshire), 85, 97
Kirk and Baum, coal mine operators, 131-32
Kuffler, Abraham, 23

Lambeth (England), 7, 38, 41-42, 45, 48, 50-52, 58, 61, 101
Lancashire (England), 100
lead, 42, 48, 98
Lehigh valley (Pennsylvania), 147
Leibniz, Baron von, 58

Wilkinson, William, 97-98
windmills, 19, 33, 48
Winlaton (Newcastle), 76
Woodbury, Robert S., 4
Woolwich (England), 7, 21-24, 27
Worcester, Marquis of, 5, 38, 40-
41, 44-47, 50, 52; *Century of Innovations*, 45-46, 50

Worcestershire (England), 38, 70,
120-21

Yorkshire (England), 100
Youlden, William, 42

Zonca, Vittorio, 32